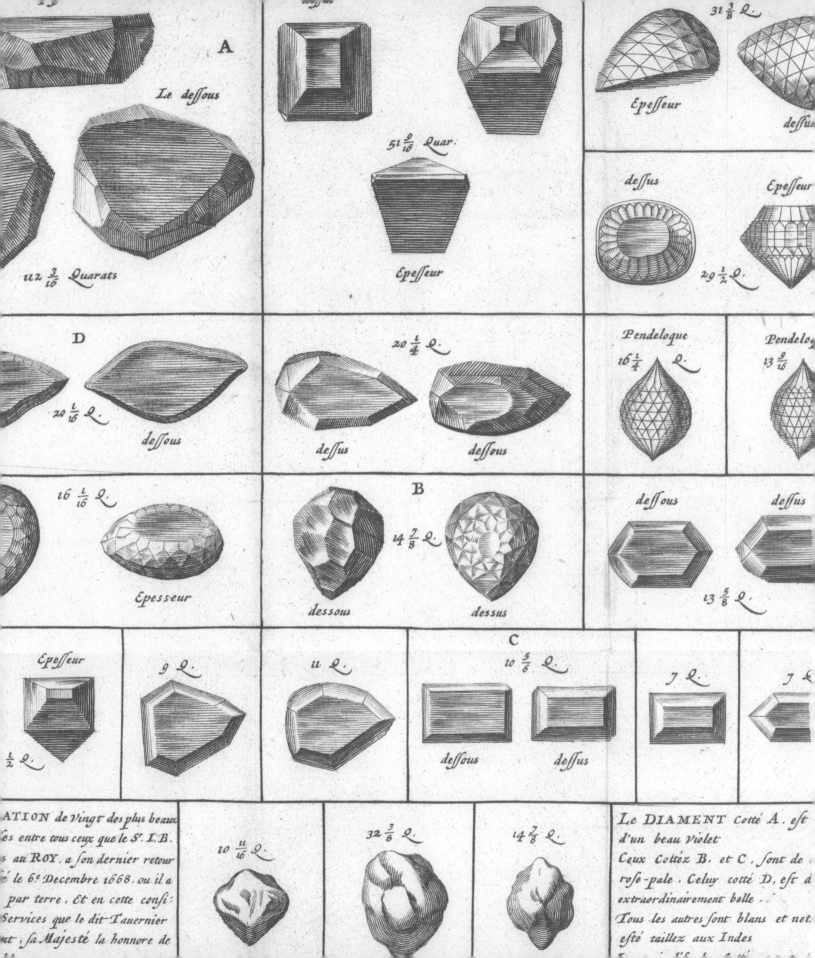

Le deſſous

A

112 $\frac{3}{16}$ Quarats

51 $\frac{9}{16}$ Quar:

Epeſſeur

31 $\frac{3}{8}$ Q.

Epeſſeur

deſſus

deſſus

Epeſſeur

29 $\frac{1}{2}$ Q.

D

20 $\frac{1}{16}$ Q.

deſſous

20 $\frac{1}{4}$ Q.

deſſus

deſſous

Pendeloque

16 $\frac{1}{4}$ Q.

Pendelo,

13 $\frac{9}{16}$ Q.

16 $\frac{1}{16}$ Q.

Epeſſeur

B

14 $\frac{7}{8}$ Q.

deſſous

deſſus

deſſous

deſſus

13 $\frac{5}{8}$ Q.

Epeſſeur

$\frac{1}{2}$ Q.

9 Q.

11 Q.

C

10 $\frac{5}{6}$ Q.

deſſous

deſſus

7 Q.

7 Q

ATION de Vingt des plus beaux
ſes entre tous ceux que le S^r. L.B.
s au ROY, a ſon dernier retour
é le 6e Decembre 1668. ou il a
par terre, Et en cette conſi:
Services que le dit Tauernier
t, ſa Majeſté la honnore de

10 $\frac{11}{16}$ Q.

32 $\frac{3}{8}$ Q.

14 $\frac{7}{8}$ Q.

Le DIAMENT cotté A. eſt
d'un beau Violet
Ceux Cottéz B. et C. ſont de
roſe-pale. Celuy cotté D. eſt d
extraordinairement belle.
Tous les autres ſont blans et net.
eſté taillez aux Indes

GEMS & CRYSTALS

FROM ONE OF THE WORLD'S GREAT COLLECTIONS

AMERICAN MUSEUM ᴼꜰ NATURAL HISTORY

GEMS & CRYSTALS

FROM ONE OF THE WORLD'S GREAT COLLECTIONS

GEORGE E. HARLOW
CURATOR OF GEMS AND MINERALS, AMERICAN MUSEUM OF NATURAL HISTORY

AND ANNA S. SOFIANIDES

PHOTOGRAPHY BY ERICA AND HAROLD VAN PELT

Sterling Signature
NEW YORK

AMERICAN MUSEUM
of NATURAL HISTORY

Sterling Signature NEW YORK

An Imprint of Sterling Publishing
1166 Avenue of the Americas
New York, NY 10036

Distributed in Canada by Sterling Publishing
c/o Canadian Manda Group, 664 Annette Street
Toronto, Ontario, Canada M6S 2C8
Distributed in the United Kingdom by GMC Distribution Services
Castle Place, 166 High Street, Lewes, East Sussex, England BN7 1XU
Distributed in Australia by Capricorn Link (Australia) Pty. Ltd.
P.O. Box 704, Windsor, NSW 2756, Australia

For information about custom editions, special sales, and premium and corporate purchases,
please contact Sterling Special Sales at 800-805-5489 or specialsales@sterlingpublishing.com.

Manufactured in China

2 4 6 8 10 9 7 5 3 1

www.sterlingpublishing.com

The American Museum of Natural History is one of the world's preeminent scientific, educational, and cultural institutions, with annual attendance of approximately 5 million. Since its founding in 1869, the Museum has pursued its mission—to discover, interpret, and share information about human cultures, the natural world, and the universe—through a broad program of scientific research, education, and exhibition. The Museum's Research Library—one of the largest natural science libraries in the world—contains extensive rare book, manuscript, photographic, and archival collections. The Museum's mineral and gem collection is one of the world's greatest, with roughly 115,000 minerals and 4,500 gems. The Museum's 45 permanent exhibition galleries include some of the world's greatest dioramas, in addition to fossil halls and the Theodore Roosevelt Memorial Hall, and its Rose Center for Earth and Space is home to the Hayden Planetarium. The Museum's collections, only a tiny fraction of which are on view, surpass 33 million specimens and artifacts. They are an invaluable resource for the Museum's 200 scientists, for graduate students in its Richard Gilder Graduate School—the Western Hemisphere's only museum-based Ph.D.-granting program—and for researchers around the world.

PREVIOUS PAGES: Photograph of the American Museum of Natural History, ca. 1902.
TABLE OF CONTENTS: Variety of gemstone crystals (see page 11).

" The love of precious stones is deeply implanted in the human heart, and the cause of this must be sought not only in their coloring and brilliancy but also in their durability. All the fair colors of flowers and foliage, and even the blue of the sky and the glory of the sunset clouds, only last for a short time, and are subject to continual change, but the sheen and coloration of precious stones are the same today as they were thousands of years ago and will be for thousands of years to come."

—George F. Kunz, *The Curious Lore of Precious Stones*, 1913

CONTENTS

Introduction

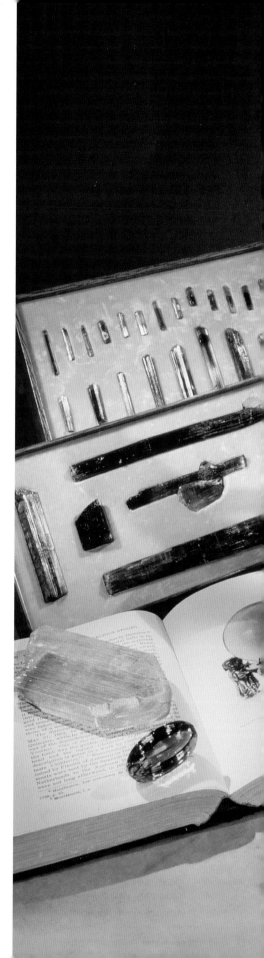

"These gems are nice, but you know they are all fakes. The real ones are kept locked up in a vault somewhere." So I recall a visitor's remark during one of my walks through the Morgan Memorial Hall of Gems at the American Museum of Natural History. This comment is not the most disarming I have heard in that great treasure chest, but it shows some of the misconceptions that exist. If this assertion were true, the Museum's insurance agents could breathe a sigh of relief, but fortunately for the public, it is not. Everything on display is real.

More interesting to me is the level of people's appreciation. "Oh, that's the Star of India. Isn't it pretty?" This response is common but a far cry from the expert's "LOOK at that pad" (pronounced *pod*), in reference to our 100-carat orangey sapphire known as the variety padparadscha. "It's FANTASTIC!" Most visitors are impressed by this gem but do not know just how special it is. One of our goals here is to give you a good look at the Museum's gems and gem crystals. The more general intent is to provide information on these that is both interesting and useful.

Minerals and gems have been part of the Museum since it opened in 1869 in the old Arsenal Building in Central Park. There was a small mineral "cabinet" to instruct the public, but it was nothing to brag about. For their growth into international prominence, the collections awaited benefactors such as Charles L. Tiffany,

Morris K. Jesup, and John Pierpont (J. P.) Morgan. George Frederick Kunz was a central figure; he was Tiffany & Co.'s gem expert from 1877 until he died in 1932, and during that time he had a profound effect on the gem industry and the Museum's gem and mineral collections.

LEFT: George Frederick Kunz, ca. 1900.

RIGHT: A historic grouping of books by George Frederick Kunz and Herbert P. Whitlock, and objects from the Museum's collection: crystals, gems, a crystal ball, and an antique crystal-measuring instrument.

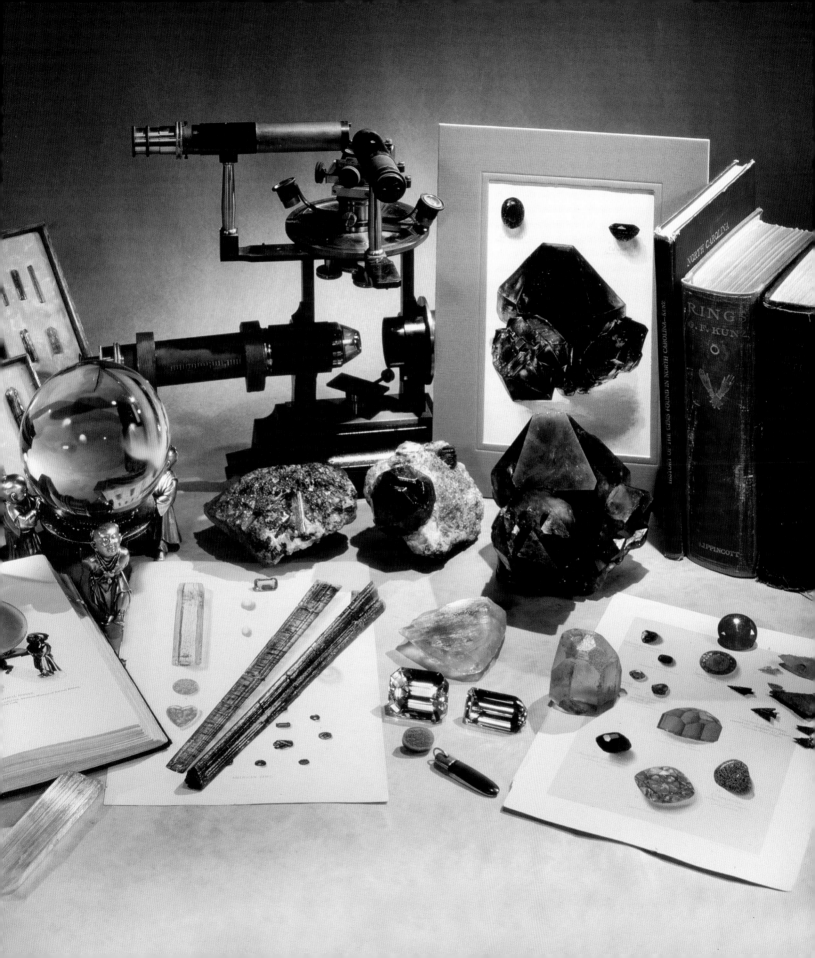

In 1889, a great Exposition Universelle was scheduled in Paris; it provided an opportunity for Tiffany & Co. to demonstrate to Europeans both American artistry in the form of jewelry and silver and North America's natural wealth through a collection of "Gems and Precious Stones" assembled by Kunz and consisting of 382 items, according the exhibition catalog. He searched the continent and gathered a formidable array of stones, crystals, pearls, and other specimens that surprised the European audience and won the collection a gold medal. Whereas much of the Tiffany jewelry was sold then and there, the "collection" was not commercial and was brought back to New York. This was to Kunz's liking, as he felt it should be kept intact, preferably coming to the American Museum of Natural History. The Museum's president, Morris K. Jesup, understood the value of the collection, but there was a problem of money—exactly $20,000. This was such a significant sum that the Museum's Board of Trustees felt inclined to question it or, at the very least, to haggle. After months of negotiations, the problem was ultimately resolved by J. P. Morgan, the banker, financier, and Museum Board member, who permitted $15,000 toward the purchase price to be "charged to his account," presumably at Tiffany & Co., which donated the remaining $5,000. Thus the Museum gained a significant gem collection in 1890, called the Tiffany Collection or Tiffany-Morgan Collection of Gems.

The same personalities and forces came together in 1900 for another Exposition Universelle in Paris. Morgan responded to the challenge and is thought to have supplied $1 million (remember, this is 1900!) for Kunz to search the world over for fabulous gems and specimens. The result was an even mightier exhibit, one that captured a grand prize. This collection, the Second Tiffany-Morgan (or Morgan-Tiffany) Collection of Gems, came directly to the Museum and included 1,453 gems—"American and foreign, cut and in their natural state," plus ninety-five pearls and shells. By

John Pierpont Morgan, ca. 1902.

1913, the gem collection contained 2,176 specimens of gemstones and 2,442 pearls and clearly laid claim to being the finest in North America, if not the world.

Morgan's interest in the gem and mineral collections continued. In 1901, he purchased for $100,000 one of the great private mineral collections created during the nineteenth century, that of Clarence S. Bement, a Philadelphia industrialist. This collection was not only superb in quality but so large that two railroad boxcars were required to bring the approximately 13,000 specimens to the Museum. This addition became the backbone of our mineral collection; many of its fine pieces are currently featured in the mineral and gem halls. Morgan's donations continued until his death in 1913.

A list of noteworthy donors to the collection would be extremely long, but I would like to mention a few more. J. P. Morgan Jr. continued his father's tradition and is responsible for many of the large fine gems, particularly a group of sapphires donated in 1927. George F. Baker, a friend of the elder Morgan, funded the creation of the Morgan Memorial Hall, which opened on the Museum's fourth floor on May 1, 1922. Kunz, who was not only responsible for Morgan's gifts but for those of many others, contributed numerous specimens and several collections to a variety of institutions. He was named honorary curator of precious stones in 1904—a title never bestowed before or since. William Boyce Thompson, the founder of Newmont Mining Corporation, provided a significant fund in 1940, the earnings from which permitted us to purchase specimens such as the Harlequin Prince black opal, a fabulous 59-carat heart-shaped morganite, and the 596-pound topaz crystal. In 1951, upon the death of Gertrude Hickman Thompson, his widow, many more magnificent gems, carvings—particularly of jade—and minerals came to the Museum.

Some have given individual stones so spectacular that each carries the donor's name: Edith Haggin DeLong (the DeLong Star Ruby), Elizabeth Cockroft Schettler (the

Schettler Emerald), and Zoe B. Armstrong (the Armstrong Diamond). Harry F. Guggenheim gave both gifts and his name to the present Hall of Minerals, which together with the Morgan Hall of Gems opened in May of 1976.

FOR MOST OF the Museum's history, the mineral and gem collections have been administered by a single curator: from 1869 to 1876 by Albert S. Bickmore, the Museum's founder; from 1876 to 1917 by Louis Pope Gratacap, who in his tenure of more than forty years established the Department of Mineralogy and procedures for managing the collection; from 1918 to 1941 by Herbert P. Whitlock, who wrote numerous books on the collections; from 1936 to 1952 by Frederick H. Pough, who concentrated on developing the gem collection; from 1953 to 1965 by Brian H. Mason, an academic geologist who developed a particular interest in meteorites; and from 1965 to 1976 by D. Vincent Manson, who devoted his energies to creating the new gem and mineral exhibition halls. The collections also have many unsung heroes—the assistants to curators. In particular, Dave Seaman cared for the collections from 1950 until he retired in 1974; Joe Peters succeeded in this role until 2002; and Jamie Newman has been in the role since that time.

In the late 1970s, the Museum recognized that one curator could not manage a department with four important collections—minerals, gems, meteorites, and rocks—and conduct the scientific research expected of all curators. In connection with the opening of new mineral, gem, and meteorite halls in 1976, the Department took a new name, Department of Mineral Sciences, and began to expand. Martin Prinz, a petrologist, became curator of the meteorite collection and chairman; and I joined the

Harry F. Guggenheim, ca. 1920.

Museum as curator of minerals and gems later that year. The Department now includes four curators who are responsible for the collections of meteorites, rocks (petrology), mineral deposits, and minerals and gems. To reflect the expanding research of the Department, the name was changed in 1995 to the Department of Earth and Planetary Sciences.

The most notorious event in the Department's history happened on October 29, 1964, when Jack (Murph the Surf) Murphy and two accomplices made a daring robbery of the old Morgan Memorial Hall, getting away with the Star of India, the DeLong Star Ruby, the Midnight Star, the Schettler Emerald, many other stones, and virtually all of the diamonds. Having seen the movie *Topkapi*, which depicts a fantastic burglary in the Topkapi Palace Museum in Istanbul, Murphy and his accomplices decided that the Morgan Hall could be entered in much the same way. Two of them hid on the top floor of the Museum while another circled outside in a getaway car. The pair lowered themselves by rope through an open window into the old Morgan Hall, where they literally raked the stones out of the cases with a glass cleaner. The only alarm in the Hall, that for the Star of India, had a dead battery. The burglars escaped easily but were so boastful about their triumph that they were quickly apprehended, and most of the stones were returned. The DeLong Star Ruby had already gone to the underworld and had to be ransomed. Thirty-five objects have never been recovered, including the uncut 15-carat Eagle Diamond from a glacial moraine near Eagle, Wisconsin, at that time the largest diamond ever found in the United States. This unique diamond crystal and others were probably cut into gems and so permanently lost—a real tragedy.

THE MINERALS AND GEMS on display number approximately 4,400, roughly 1,800 in the Morgan Hall of Gems and 2,600 in the Guggenheim Hall of Minerals. The collections actually total in

excess of 110,000 minerals and 4,500 gems; that is, a large proportion of gems are on exhibit, but only two percent of the minerals. The reason for the vastly different percentages is the nature of the materials and the multiple goals of the Museum. The gems are by definition ornamental material, increasing their value for exhibition. Mineral specimens, while sometimes manifesting spectacular crystallinity, color, or form, are frequently rather visually uninteresting—what I sometimes call "uglies"—or not even visible to the naked eye. We have such specimens on display to show a representative spectrum of the nearly 5,000 described mineral species, but the vast majority of the collection stays behind the scenes. The value in these "hidden" specimens is their record of Earth chemistry, of mineral-forming processes, and of the ways in which atoms can be arranged. The collections are a resource for scientists from universities and museums all over the world. The beautiful counterparts to uglies, the gems and gem crystals, can be even more valuable to science because of their unusual size and the perfection of crystallinity. My own scientific research in recent years has included studies of the jadeite variety of jade, an interest first stimulated by the beautiful jade objects, ruby, and peridot. However, we try to preserve the gems and beautiful crystals for both the enjoyment and edification of Museum visitors.

COLLECTIONS MUST CONTINUE to grow to stay alive. Much as the public relies on museums to display wealth once available to only a few, museums rely on the public to share such wealth. The American tradition of generosity provides the basis of most museum collections; certainly this is true for the American Museum of Natural History. The amount of donating fluctuates with the economy (and the tax code), but we should all hope that museum collections can grow and allow all of us to see the world's treasures.

The task of presenting a superb gem and gemstone collection is a formidable one and a rare opportunity. The superb color photographs in this book by the esteemed photographers Harold and Erica Van Pelt and staff photographers at the Museum speak for themselves and provide a vibrant portrayal of the collection. But the collection is much more than the sum of its images; it is a diverse resource for research and education and an archive of natural perfection. In my more than thirty-eight years as curator and my coauthor's seventeen years of work as a gemologist with the collection, we have developed a great appreciation for the gems. The text conveys some of our knowledge of them. I have focused on the distinctive properties of the gemstones and their origins, and Anna Sofianides contributed her wealth of information on their history and lore and on gem evaluation. This edition includes new images and information, as well as revisions and updates twenty-five years after the initial publication.

This book can serve as a concise visual guide to the gem collection at the American Museum of Natural History and a compendium of gem mineral information. However, no book or photograph can rival the real thing. Gems are visual delights, and seeing them firsthand is the only way to observe their character. A sparkling faceted gem comes alive with the motions of the viewer, the illumination, and the stone. The same is true for asteriated gems, cat's eyes, and opals. It is no wonder that gems have long been used for human adornment, as motion is an essence of our being. Thus I encourage you, once fortified with the images and information in our book, to visit our collection in New York City and see, while you move around the exhibit, the wonderful qualities of the gems and crystals.

George E. Harlow
February 2015

OPPOSITE AND ABOVE: Views of the Morgan Hall of Gems, ca. 2012. The large object in the picture (opposite) is a 4.5 ton pillar of azurite-malachite ore from Arizona.

ABOVE: Drawing of the Museum with the Roosevelt Memorial in front (1926) by John Russell Pope, from a hand-colored lantern slide.

RIGHT: A class from the Ethical Culture School visits the Morgan Hall of Gems in 1928. Standing at left, conducting the tour, is curator Herbert Whitlock.

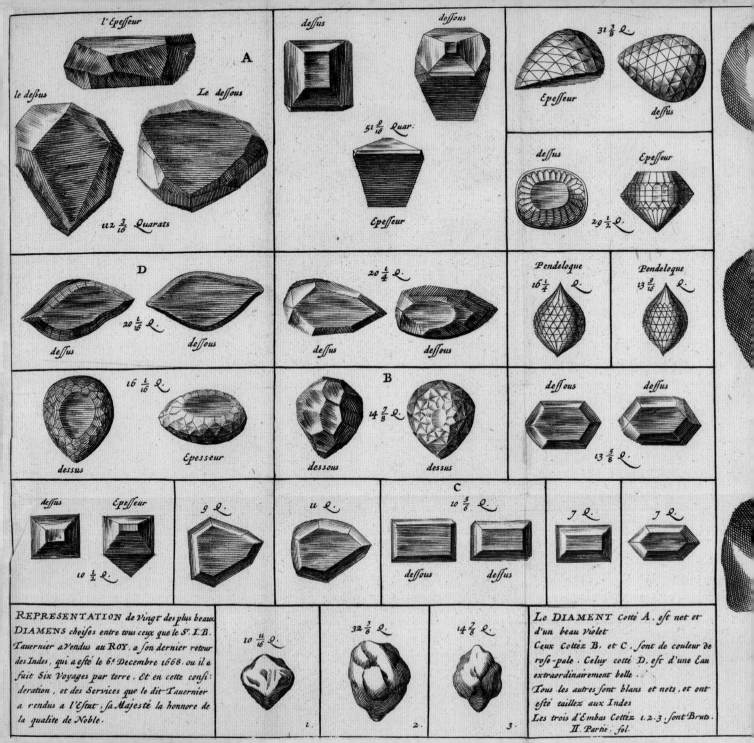

l'Epesseur

A

le dessus

Le dessous

112 3/16 Quarats

dessus   dessous

51 9/16 Quar:

Epesseur

31 3/8 Q.

Epesseur

dessus

dessus   Epesseur

29 1/2 Q.

D

20 1/16 Q.

dessus   dessous

20 1/4 Q.

dessus   dessous

Pendeloque
16 1/4 Q.

Pendeloque
13 9/16 Q.

16 1/16 Q.

dessus

Epesseur

B

4 7/8 Q.

dessous   dessus

dessous   dessus

13 5/8 Q.

dessus   Epesseur

10 1/2 Q.

9 Q.

11 Q.

C

10 5/6 Q.

dessous   dessus

7 Q.   7 Q.

REPRESENTATION de Vingt des plus beaux DIAMENS choises entre tous ceux que le S.r I.B. Tauernier a Vendus au ROY, a son dernier retour des Indes, qui a esté le 6.e Decembre 1668. ou il a fait Six Voyages par terre. Et en cette consideration, et des Services que le dit Tauernier a rendus a l'Estat, sa Majesté la honnore de la qualité de Noble.

10 11/16 Q.

32 3/8 Q.

14 7/8 Q.

1.   2.   3.

Le DIAMENT cotté A. est net et d'un beau Violet
Ceux Cottez B. et C. sont de couleur de rose-pale. Celuy cotté D. est d'une Eau extraordinairement belle.
Tous les autres sont blans et nets, et ont esté taillez aux Indes
Les trois d'Embas Cottez 1.2.3. sont Bruts.
II. Partie. fol.

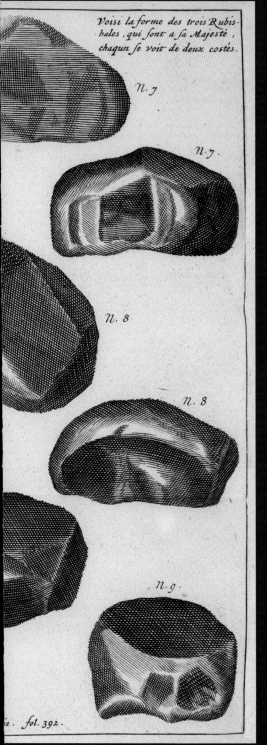

PART ONE

Voisi la forme des trois Rubis-
bales, qui sont a sa Majesté,
chaqun se voit de deux costés.

n. 7

n. 7.

n. 8

n. 8

n. 9

ie. fol. 392.

THE WORLD OF GEMS

"Twenty diamonds sold by Tavernier to Louis XIV,"
from *Les six voyages de Jean-Baptiste Tavernier*... (1678),
from the American Museum of Natural History Research Library.

TRACING THE
STORY OF GEMS

Archaeology gives the earliest picture; it attempts to tell us when and where each gemstone was used, how it was fashioned, and whether it was traded. Recorded history provides insights into the early naming, classification, and everyday significance of gems but especially the stories that are so fascinating.

Early humans were decorating themselves with shells, pieces of bone, teeth, and pebbles by at least the Upper Paleolithic period (25,000–12,000 BCE). Most of the stones used in early civilizations were opaque and soft with bright colors or

ABOVE: The gold sarcophagus of King Tutankhamen, whose mummy was buried with 143 pieces of jewelry, including such gems as carnelian, lapis lazuli, and quartz.

OPPOSITE: Physicians and gem traders, an illustration from the *Hortus sanitatis* (*Garden of Health*) by Jacob Meyenbach (1491).

beautiful patterns. The transition to harder stones appears with carnelian and rock crystal beads that were fashioned at Jarmo in Mesopotamia (Iraq) in the seventh millennium BCE. Engraved cylinder seals appeared 2,000 years later in soft stones such as steatite and marble. Their practical function was as a means of identifying goods; when they were rolled on damp clay, a unique imprint resulted. Seals represent a significant level of technical achievement, and they were also valued as adornment and possibly a symbol of status. By the late fourth millennium in the Middle East, cylinder seals were made from rock crystal quartz, a hard gemstone, in addition to the soft stones. A woman's belt from the end of the third millennium BCE, found in Harappa, an ancient center of Indus civilization, was decorated with colorful, opaque stones—red carnelian, green steatite, agate, jasper, amazonite, jade, and lapis lazuli—representing the wealth of gems then available.

Active exploitation of the lapis lazuli mines in Badakhshan, Afghanistan, and the turquoise mines on the Sinai Peninsula began around 5,000 years ago; and long-distance trade in gems developed. Lapis lazuli from Badakhshan reached Egypt before 3000 BCE and Sumer (Iraq) by 2500 BCE. China, India, Greece, and Rome received gemstones from the same source. By around the second millennium BCE, Phoenician sea merchants were trading Baltic amber on the coasts of North Africa, Turkey, Cyprus, and Greece. Spectroscopic studies of amber beads discovered in Peloponnese graves of Mycenaean Greece (1450 BCE) confirmed the assumption of their Baltic origin. Trade between Asia and Europe expanded in the fourth century BCE after the time of Alexander the Great (356–323 BCE), resulting in an increase in the number of gemstones available.

In all civilizations, magical powers have been ascribed to gems, perhaps out of a need to explain their rarity, beauty, and strangeness in a confusing world. Color played a great role in the symbolism: gold for the sun; blue for the sky,

heaven, or sea; red for blood; black for death. Colors were also associated with the planets, and astrology gems became connected here as well. Durability was also important, the unsurpassable hardness of diamond reflects the belief that it will bring its wearer strength and invincibility. Gems have served as talismans, offering protection, preserving health, and securing wealth, love, and good luck. Of the 143 pieces of jewelry discovered on the mummy of Tutankhamen (reigned 1333–1323 BCE) made of gold, carnelian, jasper, lapis lazuli, turquoise, obsidian, rock crystal, alabaster, amazonite, and

jade, a few showed no sign of wear. Given the preoccupation of ancient Egyptian culture with the afterlife, these few were fashioned as amulets to avert evil and bring good luck after death. The others were worn during his lifetime and served as both adornments and talismans. The Babylonians (in contrast to the Egyptians) were not as concerned with life after death. Many of their engraved cylinder seals may have been primarily talismans worn for protection during life.

Written records provide an extensive overview of how gems have been perceived by their owners and users. George F. Kunz, in *The Curious Lore of Precious Stones*, has documented the evidence carefully. In her *Magical Jewels of the Middle Ages and the Renaissance*, jewelry historian Joan Evans offers a wealth of information through studies and translations of ancient literature, surviving inventories of jewelry (some of which note magical powers), and even court records. In one case, she writes: "One of the counts of the indictment in 1232 of Hubert de Burgh [Chief Justiciar of England and Ireland] was that he had furtively removed from the royal treasury a gem which made its wearer invincible in battle and had bestowed it upon his sovereign's enemy Llewellyn of Wales."

The supernatural powers of gems were regarded either as intrinsic virtues of the stones themselves or were attributed to figures, sigils, or magical inscriptions engraved upon them. These virtues were enumerated in the mineralogical and medical treatises of the time, known as lapidaries.

Medicinal powers of gems were first recorded in Western literature by the Greeks. These virtues, as well as astrological symbolism, were also recorded in Arabic lapidaries and, starting in the eighth century, in European lapidaries influenced by the Arabic works. In medieval Europe, gems were commonly worn as medicinal amulets or taken as potions. Before Pope Clement VII died in 1534, he had taken as medicine powdered gems valued at 40,000 ducats. Robert Boyle (1627–91), a great advocate of experiment in natural

history and author of *Some Considerations Touching the Usefulness of Experimental Natural Philosophy*, wrote: "I think, in Prescriptions made for the poorer sort of Patients, a Physician may well substitute cheaper ingredients in the place of these precious ones, whose Virtues are not so unquestionable as their dearness."

Throughout the chapters that follow, there will be mention of the chroniclers and their lapidaries. The first important references in Western literature are from the Greek Theophrastus (ca. 372–287 BCE), the successor of Aristotle. In his book *On Stones*, the oldest surviving mineralogy textbook, he described sixteen minerals grouped as metals, earths, and stones (the last including gemstones). This classification remained unchallenged until the eighteenth century. He identified as physical properties color, transparency, luster, fracture, hardness, and weight and also noted the medicinal values of gems. A chronicler known as Damigeron (ca. second century BCE) also wrote an early lapidary in Greek, although the original is apparently lost; a portion of the text was translated into Latin (*De Virtutibus Lapidum*) somewhere between the first and sixth centuries (see Abel 1881). Pliny the Elder (23–79 CE) compiled the knowledge of his predecessors and contemporaries to produce his 37-volume *Historia Naturalis*.

Volume 37 deals with precious stones and includes "1,300 facts, romantic stories and scientific observations" about sources, mining, use, trade, and the values of gems; gem enhancements; and gem imitations. Pliny's work was influential in Europe well into the Middle Ages. Marbode, bishop of Rennes, composed his elegant lapidary in Latin hexameter in the eleventh century. Although lacking any mineralogical significance, his work is the basis of both medicinal and magical attributes that have been cited by many later writers.

Thirteenth-century works include *Steinbüch* by Volmar, as well as the important *De Mineralibus* of Albertus

Magnus (1206–80). This German philosopher noted the magnetic properties of magnetite, experimented with decomposition of arsenic minerals, and described the properties, including magical virtues, of ninety-four minerals.

Complicating our task of finding out where initial concepts came from is the fact that "borrowing" was not uncommon. Camillus Leonardus's *Speculum Lapidum*, printed in Venice in 1502, was literally translated into Italian and republished as *Trattato delle Gemme* by Ludovico Dolce later in the same century. Another sixteenth-century work, more important than that of Leonardus, is *De Gemmis et Coloribus* by Girolamo Cardano, published in 1550–87 (many volumes). Anselmus Boetius de Boodt, court physician to Holy Roman Emperor Rudolf II, wrote *Gemmarum et Lapidum Historia* (1609); in his extensive work, he provided descriptions of gems and reports on their virtues, although we find the beginnings of doubts as to the infallible powers of gems.

In addition to traditional lapidaries, travel books—such as those by Marco Polo in the thirteenth century and Jean-Baptiste Tavernier's *Les Six Voyages ... en Turquie, en Perse, et aux Indes* in 1676—documented information about the use of gems and their sources, particularly diamonds in India. Garcia de Orta (1565), Portuguese physician to the viceroy of Goa in India, described diamond mines there, observed mining practices, and reported gems' virtues. He flatly denied a then-commonplace belief that the diamond was poison, having seen workers swallow the gem in order to smuggle it.

With the development of empiricism and scientific inquiry in the seventeenth and eighteenth centuries, the study of gemstones as manifestations of a strictly physical universe began. Concepts of chemistry, optics, and crystallography developed along with a desire to categorize so that definitions and tests could begin to differentiate among all objects.

Today we view gems from a very different perspective from that of a few hundred years ago, but we still have much to learn. Color, the great deceiver in the transparent stones, is still a subject filled with questions about specific causes in each gemstone. New gemstones and new treatments (with chemicals and heat) of the old are discovered and add to the diversity. The beauty of the challenge is the gems themselves; they offer a wonderfully exciting stimulus for exploring nature.

ABOVE: The cover of *Gemmarum et Lapidum Historia* (1609) by Anselmus Boetius de Boodt.

RIGHT: Frontispiece from *Les six voyages de Jean-Baptiste Tavernier...*, depicting Europeans meeting with Indian diamond miners.

WHAT IS A GEM?

The purpose of this chapter is to answer this question and to discuss the attributes that distinguish gems. The rest of the book examines the gemstones by mineral group; it starts with the traditional "precious" stones, then moves through the colored stones, ending with organic gems, rare and unusual gemstones, and

ABOVE: Myriad colored stones, including topaz, amethyst, aquamarine, morganite, chrysoberyl, peridot, smoky quartz, citrine, calcite, rhodochrosite, kunzite, and fluorite, varying from 14.92 to 454 carats.

WHAT IS A GEM?

ornamental material. However—aside from the "precious" four: diamond, ruby, sapphire, and emerald—there is neither a system nor agreement on how to order gems, because beauty, their hallmark, is a matter of taste and culture. Pearls, jade, and opal are highly regarded, but ranking them—who knows? Moreover, tastes and availability change with time; today's ranking could be noticeably different in a decade. You undoubtedly have your own favorites, but browse through all the chapters for some beautiful surprises.

Getting back to the title question: to us, a gem is a gemstone that has been fashioned—cut, shaped, and/or polished—to enhance its natural beauty. The gemstone is the raw material or "rough"; the gem is the finished product. Most gemstones are minerals, but some are rocks, and a few are the organic products of once-living animals or plants. A gem ruby is created from a piece of the mineral corundum; lapis and jade are rocks; and pearls, amber, and jet are organic products.

What Gems Are Made Of

Rocks and minerals represent two different levels of organization of matter. Atoms are the building blocks for minerals; minerals are the building blocks for rocks; and rocks make up the solid Earth. Rocks are physical assemblages of pieces of minerals. The boundaries between all the grains tend to make rocks opaque, so it is usually their color or occasionally their toughness or translucency, or a combination of these, that makes them attractive as gemstones. Minerals are more frequently valued as gemstones because they can be transparent and fashioned into glittering gems. The fact that minerals are crystalline allows them to form large transparent gemstones—crystals—and have interesting optical properties.

Being crystalline means that a mineral's constituent atoms—of one or more of the ninety-two naturally occurring chemical elements—are organized in a regular geometric pattern that repeats in three dimensions to form a solid body called a crystal. One usually recognizes a crystal by its natural flat surfaces called faces. Crystals can be invisibly small or larger than automobiles, but a crystal's properties and atomic organization are the same throughout the entire body. The geometric arrangement of atoms is called the crystal structure. The atoms are linked by chemical bonds—electrical forces and electron interactions between atoms—that literally hold the mineral together and produce its properties. A mineral also has a chemical composition that can be defined

A quartz display case at the American Museum of Natural History, containing amethyst, citrine, smoky quartz, rock crystal, rose quartz, rutilated quartz, and green quartz.

7

within well-constrained limits, usually by a relatively simple chemical formula. Corundum is natural aluminum oxide, Al_2O_3. Each of the roughly 5,000 mineral species is defined by its chemical composition and its crystal structure.

A well-formed crystal is the external expression of the symmetry of the repeating arrangement of atoms within. The way in which shapes are repeated determines the type of symmetry. A crystal with faces that repeat across a plane (a mirror), around an axis (rotation), or through a point (inversion) contains one (or more) elements of these three basic kinds of symmetry. For example, a hexagonal green beryl crystal rotated 60 degrees around its center (axis) appears unmoved; it is repeated by six-fold rotation. Some minerals have no symmetry, and others have many symmetry elements. The symmetry a crystal can possess is constrained by the ways groups of atoms can be repeated in space to create a solid crystal. There are only seven basic systems of crystal symmetry, which are shown diagrammatically on the opposite page. All minerals belong to one of these seven crystal systems. The crystal symmetry is not only helpful in identification, as with the hexagonal beryl, but properties such as hardness and color can vary with direction in the crystal, depending upon its symmetry. Thus, a gem is cut from an individual crystal to take advantage of not only its uniformity as a transparent, cohesive object but, possibly, its special directionally dependent properties.

Mineral crystals are created by growth from a nucleus, some speck or mineral surface, by additions of successive layers of atoms on the crystal's outer surface. Some minerals, particularly tourmaline among gemstones, manifest this layered growth with concentrically color-banded crystals. Growth occurs when the temperature, pressure, and chemical environment are favorable; but most mineral crystals, such as the grains in rocks, do not have symmetric form with flat faces because the crystals grow into or against one another. Well-formed crystals are rare because they need not only appropriate and sustained growing conditions but also a space in which to grow, such as a cavity, where growth will be unimpeded. Even then, perfectly symmetrical crystals are rare.

The term *habit* refers to the actual shape of a mineral crystal or aggregate. For example, a common habit for crystals is the prism, a form with parallel sides that looks like an extruded polygon. Most pencils, like beryl crystals, have the form of a hexagonal prism. Some minerals, such as chrysoberyl, grow as multiple crystals, known as *twins*. Two or more crystals develop in intimate contact, as an "intergrowth," and give the appearance of a single crystal with more symmetry than the mineral possesses, such as manifesting a six-fold rotation axis where there is none. Other minerals form fine-scale, rocklike aggregates, such as nephrite jade or the jasper form of quartz. A mineral's habit may vary, depending on the conditions during crystal growth.

Minerals are said to form a *group* if they share the same crystal structure—the atoms have the same arrangement, but the chemical elements vary. Garnet is a mineral group with many members. Two gemstone members in the garnet group are spessartine, $Mn_3Al_2Si_3O_{12}$, and almandine, $Fe_3Al_2Si_3O_{12}$. Because the sizes of manganese (Mn) and iron (Fe) atoms are similar, a continuum of mineral compositions between spessartine and almandine is possible. This phenomenon, called *solid solution*, is important to diversity (and complexity) in gemstones like the garnets. Mineral names within solid solution series were once common, but, today, whichever limiting composition is dominant defines the name; a garnet with 55 molecular percent almandine and 45 percent other species is an almandine.

A color or compositional variant of a mineral is termed a *variety*; e.g., emerald is a green variety of beryl. Many of the appellations associated with gems are variety rather than mineral names.

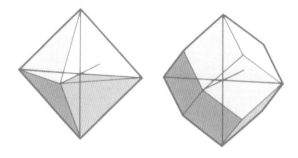

CUBIC—Diamond and Garnet

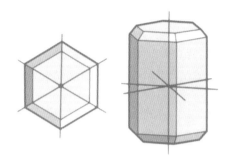

HEXAGONAL—Beryl

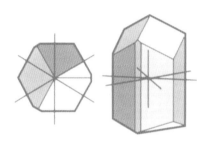

TRIGONAL—Elbaite (Tourmaline)

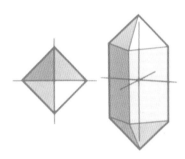

TETRAGONAL—Zircon

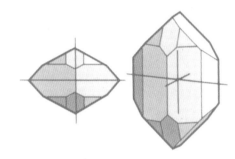

ORTHORHOMBIC—Topaz

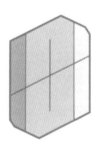
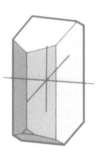

MONOCLINIC—Orthoclase (Feldspar)

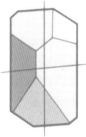
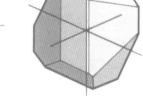

TRICLINIC—Amblygonite

CRYSTAL SYMMETRY SYSTEMS | The combination of symmetry and repetitive organization of atoms produces only seven basic geometries for defining crystals. The diagrams show a crystal and axes for gem minerals in each of the seven crystal systems.

Important Gem Properties

The most significant characteristic of a gemstone is its visual beauty; after this come durability and rarity, but without beauty, the others mean little. The beauty of ruby, emerald, and turquoise lies in magnificent, intense colors, while that associated with diamond is the complete absence of color combined with high brilliance. Flawless transparency is critical for the beauty of diamond, aquamarine, and topaz, while inclusions account for the presence of a star in ruby and sapphire and the cat's eye. The lively play of color in opal and the pleasing iridescence in labradorite and moonstone are unique for these gems, as are the numerous patterns in agate. To understand what gives a gem its most important characteristic, visual beauty, we need to examine the ways in which a mineral interacts with light.

Light, Vision, and Color

The source of color is light, its interaction with an object, and our ability to perceive the result. The color we see is the light that is reflected or transmitted and not absorbed. The causes of color in gemstones are many and varied.

If a mineral's color is inherent, it is called *idiochromatic*, "self-colored." Malachite, copper carbonate, is always green because copper causes the color and is intrinsic to the mineral.

Minerals that owe their color to physical effects, such as internal boundaries and contaminants, are called *pseudochromatic*, "false-colored." Jasper, a form of quartz with extremely fine grain size, can contain small particles of iron oxide (hematite) that make it brick red. Physical scattering of light, described a little later, produces the play of colors in precious opal.

Allochromatic, "other-colored," minerals are generally colorless and transparent in their pure state but develop color with minor changes in crystal composition or from structural imperfections. Such gemstones are the most numerous, intriguing, and difficult to identify by color alone. Substitution of certain transition elements for aluminum in corundum yields a variety of colors: some iron and titanium causes blue sapphire; a little iron alone results in yellow sapphire; a little chromium produces ruby. (Transition elements are chemical elements in the middle of the periodic table whose electron energy transitions can be stimulated by visible light, thus yielding color.) The same element can result in different colors; a minor substitution of chromium for aluminum in colorless beryl produces the spectacular emerald. Other transition elements important in causing color are manganese, copper, and vanadium.

Damage and/or mistakes in the crystal can cause color. Smoky quartz is the result of radiation damage to the crystal related to substitutional impurities. It can be produced by naturally occurring radioactive minerals adjacent to a quartz crystal or by bombardment with subatomic particles from a nuclear reactor. The general term for this type of color source is known as a *color center*.

Color in some crystals changes with their orientation; the phenomenon is called *pleochroism*, "more coloring." Sapphires and rubies and pink spodumene are more deeply colored when viewed down the prism axis. Tourmaline gems can have two different colors, depending upon the direction you look through the gem or crystal. Orientation is very important to the appropriate fashioning of pleochroic gemstones.

A few gems, notably the ruby, have the property of fluorescence; they can absorb blue and ultraviolet light and reradiate some of the energy in a redder portion of the spectrum. The result is a more intense color—redder than red—an extra glow that dazzles the eye.

Gems' Sparkle

Another important pair of related optical properties of a gem is the way it reflects and refracts light. Luster is the reflection or scattering of light from a gem's surface; it can range from metallic to vitreous to resinous to earthy. High luster requires both a smooth surface and a high reflectivity. Polishing of all gems is important in part to improve luster.

Brilliance, on the other hand, is the reflection of light from inside a faceted gem. ("Life" and "liveliness" are used synonymously for brilliance.) This quality is a function of both the cut and the refractive index (R.I.). The R.I. is actually a measure of the velocity of light in the gemstone but is manifested by the degree to which light is bent when entering a substance at an angle and the critical angle at which light is reflected instead. The angles of cut, and thus

ABOVE: Gemstone crystals including tourmaline (elbaite), aquamarine, morganite, heliodor, topaz, kunzite, spodumene, and citrine.

11

a gem's proportions, are specifically gauged to the R.I. of each gemstone so that the faceted gem will reflect back from inside the light that enters. All minerals except those of the highest symmetry—cubic—actually have two or three R.I.s and are called "birefringent." Reflectivity is positively correlated with refractive index, and both increase with a substance's density. Thus it is not an accident that the fine transparent gemstones like diamond and sapphire are denser than most minerals. (Density is measured in terms of specific gravity, the weight of a substance relative to that of an equal volume of water.)

Fire develops in a gem from the phenomenon known as *dispersion*. The component colors in white light are bent to varying degrees during refraction because the velocity of light—hence refractive index—within a gem varies with the wavelength of light. The separation of colors into the rainbow when light passes through a prism is a demonstration of this fact. For two different gems of the same size and cut, the one with greater dispersion will display a better spectrum of colors, or fire; the diamond will have better fire than the quartz gem. In colored gemstones, dispersion is usually masked by the predominant color, so that this property becomes unimportant.

Scattering of Light

Small to submicroscopic features can produce some surprising visual effects in gemstones. Reflections from parallel layers of transparent materials cause pearly luster. The pearl is built up of concentric layers and gives this luster its name. The cat's eye effect, *chatoyancy* (a literal translation from French), is a band of reflected light that appears in certain gemstones polished with a rounded surface (*en cabochon*). The cat's eye is produced by many straight parallel fibrous inclusions that scatter light perpendicular to their long direction. In corundum, three directions of needles can occur, yielding multiple chatoyancy in the form of a six-rayed star. This is called *asterism*. The star is visible only when the stone is viewed down the axis of intersection of the inclusions.

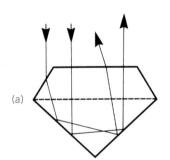
(a)

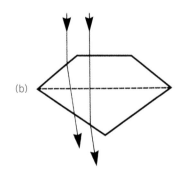
(b)

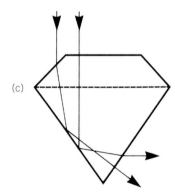
(c)

REFRACTIVE INDEX AND BRILLIANCE | A gem properly proportioned with respect to its refractive index will reflect back the light that enters it (a), yielding maximum brilliance. Light "leaks" out of a gem cut too shallow (b) or too deep (c), diminishing brilliance.

LIGHT DISPERSION | White light separates into component colors on passing through (above) a glass prism and (below) a colorless gem due to dispersion—the result for the gem is "fire."

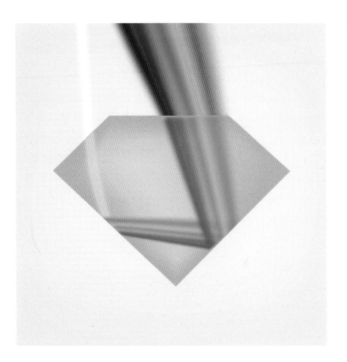

Scattering from very small features often imparts colors. In moonstone, thin layers and small elliptical bodies scatter blue light most effectively and yield the characteristic pale blue sheen. Small particles arranged in a periodic pattern will scatter individual colors by optical diffraction, which is observed commonly in bird feathers and butterfly wings. The best example in gemstones is precious opal. A number of other similar scattering phenomena produce colors in stones like fire agate.

Durability

Durability, the most important physical attribute in gauging a gemstone's merit, has three aspects: hardness, toughness, and stability. Hardness is the resistance to being scratched and is literally a measure of the strength of the chemical bonds in a substance. In 1822, German mineralogist Friedrich Mohs proposed a scale of hardness consisting of 10 minerals ranked in order of their ability to scratch one another; 1 is the softest, and 10 is the hardest. The scale is relatively linear; that is, each mineral is nearly one value of hardness greater than the previous one. Diamond, 10, is anomalous; it should have a value more like 80 to show its hardness relative to the others. Diamond is held together with extremely strong chemical bonding.

Quartz, with Mohs hardness of 7, is a common component of dust, so that gemstones softer than 7 are subject to scratching, particularly in rings, where abrasion is commonplace.

Toughness is a gem's resistance to cracking, chipping, and actually breaking. A chief threat to crystals is planes of weakness, representing directions in the crystal structure with relatively fewer or weaker chemical bonds. The result is *cleavage*—splitting along a plane. Diamond, the hardest mineral, lacks toughness because of its octahedral cleavage planes. Most diamonds in engagement settings will show small chipped corners after years of exposure to everyday wear and tear. Topaz, with a hardness of 8, also lacks toughness. It has one perfect cleavage and therefore is difficult to facet. Some gemstones fracture easily as a result of internal stress, which lowers their strength. Both opals and obsidian can chip easily due to physical or thermal shock. Nephrite jade, with a hardness between 6 and 6.5, is the toughest of gemstones. With a strong interlocking network of fibrous crystals, it can be fashioned into the most intricate shapes. Another tough gem is the pearl; one will not break if dropped on a hard floor, although the gem's hardness is only about 3.

Stability, the resistance to chemical or structural change from deteriorating forces, is an important factor in a gemstone's durability. Opals contain water, and some lose it in dry air; the result is cracks, or crazing, from loss of volume. Pearls are damaged by acids, alcohol, and perfume. Porous gemstones like turquoise can pick up oils and coloring from the skin. The color of some kunzite and amethyst fades on exposure to sunlight. However, the majority of gems are stable in most conditions the wearer is likely to place them.

MOHS HARDNESS SCALE

1. Talc
2. Gypsum
3. Calcite
4. Fluorite
5. Apatite
6. Orthoclase
7. Quartz
8. Topaz
9. Corundum
10. Diamond

Where Gems Come From

Gemstones are uncommon in the mineral kingdom and require unusual geological conditions for their creation. They can form at various depths within Earth's crust or even below in the mantle. All three classes of rock-forming environments—igneous, metamorphic, and sedimentary—produce gemstones, although the first two are predominant. Important gemstone sources, or occurrences, are gem pegmatites. Crystals that are measured in inches and feet occur in granitic pegmatites. They crystallize from the molten rock, magma, as the final step after a large quantity of granitic rock has already solidified. The residual magma becomes rich in volatiles such as fluorine, boron, lithium, beryllium, and water. The volatiles promote the growth of large crystals and are also components of aquamarine, tourmalines, and topaz, for example. The pegmatites of Minas Gerais in Brazil, the Ural Mountains of Russia, Madagascar, and San Diego County, California, are remarkable for their gem-quality crystals.

Another environment in which gemstones are found that has nothing to do with their original formation is *placers* or alluvial (river) deposits. Minerals released by weathering of rocks exposed at Earth's surface wash into rivers (and onto beaches), where they concentrate as gravels; less durable minerals break up and wash away. Dense minerals are most effectively concentrated in this way. Alluvial gemstones are often superior to those found in solid rock because only the strongest, most perfect specimens survive the abrasive transport.

OPPOSITE: The pegmatite dike at the Himalaya Mine, San Diego County, California, is famous for its production of elbaite specimens and for the classic zoning from the dike wall into the core of the dike.

15

Ancient and modern carvings
from China and Japan, including
rutilated Brazilian quartz sphere
and fu dog; nephrite pi disk;
lapis junk; serpentine box; yellow
serpentine vase; carnelian vase;
agate guppy; malachite vase; and
an aquamarine-like glass vase.

Gems and the Market

Practical questions also determine which minerals and rocks can be used as gems. Does a gemstone occur in pieces of sufficient size to fashion a reasonable gem? Many minerals could be used as gems if they occurred in clear crystals weighing several carats. Olivine is a common mineral, but geological occurrences of the green peridot crystals of adequate size are rare.

To be commercial, the gemstone must be sufficiently abundant to underwrite the cost of developing a demand. Inadequate supply leads to unviable economics. Today the issue is often the depletion of sources. Alexandrites were never plentiful, but now the supply is so scarce that few examples ever reach the mass market. Consequently, even with a spectacular, though short-lived, find at Lavra de Hematita in Brazil in 1987, alexandrite is almost unknown except by collectors and gem experts.

Is a gemstone sufficiently rare to have status? Because gems are often the hallmark of social status and wealth, a degree of rarity is important. A gemstone may not lose its appeal with overabundance, but its monetary value will certainly be affected.

CARAT, THE STANDARD UNIT OF MASS FOR GEMS

The abbreviation for *carat* is *ct*.

    1 ct. = 0.2 grams
    5 cts. = 1.0 grams = 0.035 ounce (avoirdupois)
    141.75 cts. = 28.35 grams = 1 ounce

Do not confuse *carat* with *karat*, the unit of measure of gold purity. Both terms probably originate from the Middle Eastern word for the seed of the carob tree (Arabic *quirat*). The seeds have remarkably uniform weight and were used to balance the scales in the ancient markets.

Evaluating Gems

Evaluation of gems is a search for and comparison with perfection. There is a great range in qualities of some properties in each gemstone and no absolute code by which to compare different gems.

Color. A fine colored gem must have a good depth of color, not so pale as to be uncertain and not so deep as to appear black. Different color saturation can mean a remarkable difference in the price of two gems. The color should be uniform, not blotchy or strongly zoned. For some gemstones like amethyst, finding uniformity is the most serious problem in locating a fine sample. In multicolored gemstones, sharp, straight boundaries to color change are important rather than uniformity.

Clarity. This characteristic is important in most gems, less so in others. Two varieties of beryl have very different limits to what is acceptable. A fine aquamarine should be virtually without flaw, but a fine emerald will almost certainly have a few small inclusions. In fact, an emerald free of inclusions is suspected of being synthetic. Flawless transparency— freedom from inclusions and cracks—is critical to the beauty of gems like diamond and topaz.

Weight. The weight of a gem is always a factor in determining price. The value of ruby, diamond, and emerald increases dramatically with weight because large crystals are rare. This increase is an actual rise in the price per carat. Topaz, aquamarine, and rock crystal increase far less in value with increasing size, because large crystals of these gemstones are relatively abundant. When a stone exceeds the size that can be readily set in a piece of jewelry, its unit price plateaus or even begins to fall.

Cutting and Polishing. For an opaque gemstone, only the surface properties are important. Such stones are rarely faceted and are more frequently polished to obtain a smooth, rounded surface. The cabochon ("bald head"), a rounded top usually with a flat base, is used for translucent and opaque gemstones and gems displaying optical phenomena such as chatoyancy, asterism, and play of color.

Transparent gemstones are faceted. The process consists of cutting with an abrasive (usually diamond) saw, grinding with abrasives, and polishing facets. Cut also means the shape or style in which a gem is fashioned. Faceting and proper proportioning are essential for revealing the full beauty of transparent minerals, particularly a diamond's fire. Diamond faceting first appeared in the fourteenth century, but intense study of methods was stimulated by the great nineteenth-century discoveries of diamonds in South Africa. We now know the exact angles that must be present between facets to cause all light incident on the gem to be completely reflected for maximum brilliance. The round brilliant cut with its modifications (oval, pear, marquise, and heart) and the step (emerald) cut are the most popular. New cut styles, particularly for diamond, such as princess, trilliant, and radiant, have been developed both for beauty and marketing.

The quality of cutting and polishing is another factor in the evaluation of all gems, although particularly significant for diamond. Many dealers will buy poorly faceted or proportioned stones and have them recut with reduction in weight but dramatic increase in value.

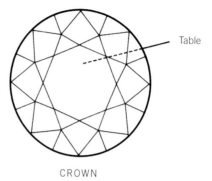

CROWN

Table

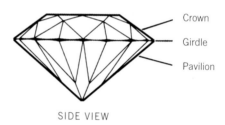

SIDE VIEW

Crown
Girdle
Pavilion

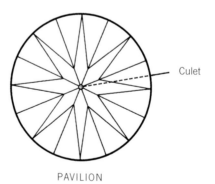

PAVILION

Culet

FACETS | The names for the various portions and facets of a cut stone diagrammed for a round brilliant.

Gem Enhancement. During recent decades, enhancement of gemstones by chemical and physical means has become very common. Irradiation is being used to enhance or change the color of many stones. For some gems, chemical treatment or impregnation is used; procedures include bleaching, oiling, waxing, plastic impregnations, and dyeing. Heat treatment improves the color and clarity of some gems and can be so intense that it pushes the boundary between natural and synthetic material. Infusing gems such as sapphire with an element at extreme temperature will change color. Not all types of treatment can be detected at present. In 1989, a resolution was adopted by the members of the International Colored Gemstone Association for disclosure of gemstone enhancement upon request by customers. The American Gemstone Trade Association has published guidelines and coding for gemstone enhancement and is committed to a policy of disclosure by its members in the gem trade.

Gem Substitutes. Substitutes are substances that so resemble a gemstone's properties that mistaken identity can occur. A relatively inexpensive gemstone may be substituted for a more valuable one, such as citrine quartz substituted for precious topaz. This practice is fraudulent.

Man-made substitutes fall into two categories: synthetics and simulants. A synthetic is the exact same substance as the natural mineral but grown in the laboratory. Synthetic gemstones have been produced commercially since the end of the nineteenth century; the first was synthetic ruby. Initially, there was fear that less expensive synthetics would dilute the market for natural stones and, thus, reduce

Pear-Shaped Brilliant Cut

Oval Brilliant Cut

Marquise or Navette Brilliant Cut

Heart-Shaped Brilliant Cut

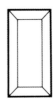

Emerald Cut

Baguette Cut

GEM CUTS | Some popular gem cuts, viewed from the table (top) and pavilion (bottom).

the latter's value; this has never occurred. Distinguishing the natural from the synthetic gem is not always easy. Natural stones usually contain some inclusions that aid identification, whereas synthetics sometimes manifest evidence, such as lines and bubbles, of their synthesis. Very sophisticated techniques may be required to differentiate the nearly perfect natural gem and the synthetic. Gem-testing laboratories have been forced to acquire the latest scientific equipment for gem authentication to keep up with the evolution of synthetics, simulants, and enhancements.

Simulants usually have no natural counterpart but have optical properties that closely resemble those of a natural gem. Cubic zirconia—zirconium oxide—is so inexpensive, so readily available, and so resemblant of diamond that it has replaced virtually all previously used diamond simulants. More recently, moissanite—silicon carbide—has entered the marketplace; however, synthetic diamond has become a bigger player as a substitute for natural diamond.

Imitation gems have been in use since antiquity. They may be glass, plastic, or composite stones consisting of two (doublets) or three (triplets) parts. These parts may be genuine, imitation, or both. A clever imitation of imperial jade is a triplet made of a hollow, colorless jadeite cabochon filled with a green jellylike substance cemented to a flat jadeite back. The color is magnificent but eventually disappears when the green substance dries. Opal is commonly made in doublets and triplets that utilize thin seams of the gem material and at the same time protect this fragile gem. Except in the instance of opal, composite stones generally have been replaced by synthetics.

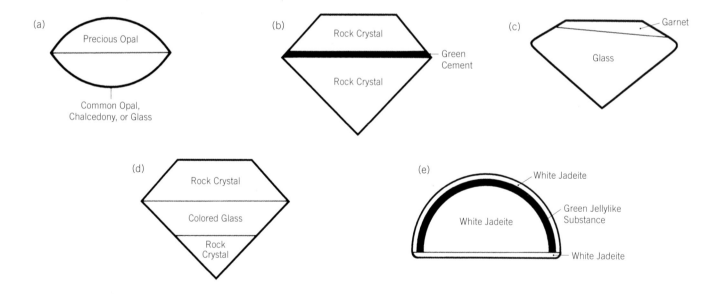

COMPOSITE GEMS | Several examples of composite gems: (a) opal doublet; (b) soudé emerald doublet; (c) garnet doublet; (d) rock crystal triplet; and (e) jadeite triplet.

PART TWO

GALLERY OF GEMS & CRYSTALS

The Morgan Hall of Gems as it looked in 1929.

DIAMOND

Diamond is the ultimate example of a gemstone with dual values. On the one hand, it is the peerless model for the colorless gem with its superior brilliance and fire and, on the other, it is the hardest of all substances; only diamond abrasive can fashion a diamond. Diamond as an ancient gem with mystical properties is tied to India, its earliest source. There its hardness was appreciated, but the wellspring for diamond's value was the purity and power that were believed to derive from the crystal form and optical properties—the ideal stone was a clear, transparent, fiery octahedron. The octahedron is the only crystal habit of diamond that manifests the "dazzling division of colors" that epitomizes the mineral's ideal high dispersion. Although diamond octahedra are not uncommon, ones that are sufficiently perfect to show fire are a rarity. This problem of scarcity in obtaining "perfection" might have been solved by reshaping (cleaving and polishing) poorly formed diamonds to bring out their sparkle and improve octahedral form. However, any alteration of a diamond was taboo, destined to destroy its mystical properties. It is a wonder that the connection of fiery optics to the popularly perceived magical virtues was ever made.

DIAMOND DATA*

Carbon: C

Crystal symmetry: Cubic

Cleavage: Perfect in four directions, defining an octahedron

Hardness: 10

Specific gravity: 3.514

R.I.: 2.417 (high), superb brilliance

Dispersion: High, superb fire

*See pages 6–19—"What Is a Gem?"— for explanation of data and definition of terms.

OPPOSITE: Naturally colored diamonds from the Aurora collection from sources around the world, from 0.75 to 2.16 cts. (formerly on loan from Aurora Gems, Inc.).

high dispersion. These properties give diamonds their superior brilliance and fire.

We usually think of diamonds as colorless gems, but most are slightly to noticeably yellowish. Intense, attractive colors like canary yellow, pink, green, blue, purplish, and the rare red are called "fancy" colors. Color is due to the presence and distribution of minor amounts of nitrogen (most yellows) or boron (blues) in diamond crystals. Defects (missing atoms and layering mistakes) and radiation damage also contribute to color; several colors can be produced by artificial irradiation, sometimes followed by heat treatment. A bluish daylight fluorescence is relatively common in diamonds, and fancy colored stones are renowned for their spectacular range of fluorescence in ultraviolet light.

Chemically diamond is pure carbon, the same as graphite, a soft mineral used in pencils. The dramatically different properties of the two minerals are caused by the chemical bonding of carbon atoms—all strong bonds in diamond, some weak ones in graphite. Diamonds are extremely stable under ordinary conditions, but, unlike other crystalline gemstones, when heated red-hot in air, diamonds will form carbon dioxide and literally vanish.

ABOVE: Single and twinned diamond crystals; maximum dimension is 1 cm (⅜ in).

Properties

A diamond's hardness is produced by the extremely strong chemical bonds each carbon atom extends to adjacent neighbors. This configuration yields a crystal structure of high symmetry as expressed by octahedron- and cube-shaped crystals. However, one direction in the structure is crossed by relatively few bonds, which results in a perfect cleavage plane that repeats via symmetry to define an octahedron. Diamonds are routinely split along this cleavage to form smaller pieces for cutting gems. Cleavage is also diamond's main failing as a gemstone—a faceted stone can break or chip. Other results of diamond's strong and compact structure are a relatively high density, high refractive index, and

Historic Notes

Diamonds were revered and highly valued as talismans at least as far back as 800 BCE in India, and for more than twenty-five centuries that country was the only known supplier of the gemstone. Diamonds and their reputation for metaphysical powers arrived in Rome around the first century BCE. Pliny describes this substance that will scratch all others, confirming the name for the hardest of substances—*adamas*. The word is derived from Greek, meaning "untamable" or "indomitable," and well suits diamond's hardness.

During the first century CE, prominent Romans wore uncut diamonds set in gold rings as talismans, and leading figures were awarded diamonds by the emperor; specimens of the period are yellowish or brownish, possibly indicating that India's best stones were not exported and almost certainly that the stones available were worn for their powers rather than their beauty.

According to Pliny, diamonds were known only to kings; indeed, very few kings possessed them even as late as the thirteenth century. Louis IX of France (1214–70) issued an order forbidding women, including queens and princesses, to wear them. Agnes Sorel, the mistress of King Charles VII (1403–61), dared to wear diamonds, and she is credited with popularizing them in the French court. During the second half of the century, the gem was more frequently worn but only by royalty. Although wedding rings date to the second century BCE, and gems in them are mentioned in Dante's *Purgatorio* (ca. 1310), the first recorded diamond betrothal ring was given to Mary of Burgundy by Hapsburg Emperor Maximilian I in 1477.

RIGHT: Portrait of Emperor Maximilian and Mary of Burgundy with their son and grandchildren, by Bernhard Strigel, ca. 1516; Mary received the first known diamond engagement ring from the emperor.

Louis XIV of France (1638–1715), the Sun King, collected many fine-quality diamonds. Jean-Baptiste Tavernier, jeweler, merchant, and traveler, was influential in bringing diamonds to the court's attention. He made six trips to the Orient, visited Indian diamond mines, and brought back fabulous stones.

Cutting began in the fourteenth century in India and Europe—it is not clear who was first—and diamond became a gem in the modern sense, treasured for its sparkling beauty. However, cutting styles evolved over centuries, with the earliest gems resembling octahedra and appearing black with white edges in paintings of the Renaissance.

During the eighteenth century, diamonds became gems "par excellence," although exclusively for the superrich. Monarchs maneuvered for possession of them; the histories

of famous and notable diamonds read like adventure stories and fairy tales.

Diamond production in India began to wane in the eighteenth century, but in 1725, diamonds were discovered at Tejuco in Brazil, and the town was renamed Diamantina. Other deposits were found, and Brazil became the world's major supplier. Toward the end of the nineteenth century, Brazilian production waned, and a series of events changed the diamond world dramatically.

Erasmus Jacobs's son found a diamond the size of a marble on the De Kalk farm near the Orange River in South Africa in 1866. It was the first diamond to be found in South Africa and was appropriately named "Eureka." After several other discoveries, a diamond rush began. In 1871,

THE KOH-I-NOOR

This magnificent gem has the longest history of all the famous diamonds. In 1304, it was in the possession of the rajah of Malwa. In 1526, it fell into the hands of the founder of the Mogul dynasty and was passed down the line to all the Great Moguls until 1739, when Nadir Shah of Persia invaded India. All of the treasures of the Moguls fell into his hands, except the great diamond. He was told that the emperor had the stone hidden in his turban. So, in accordance with local custom, he invited his vanquished opponent to a feast where turbans would be exchanged. Later, in private, Nadir Shah unrolled the turban and reportedly exclaimed, *"Koh-i-Noor!"* (Mountain of Light) when the gem tumbled to the floor.

Later, the stone was taken to Afghanistan by General Ahmad Abdali, who had served Nadir Shah. One of the sons of Ahmad Abdali (Ahmad Shah as ruler) fled to the Sikh maharajah Rangit Singh in Lahore in 1810, forfeiting the diamond soon after. After the Sikh wars, the gem was taken by the East India Company as part of the indemnity levied in 1849 and was subsequently presented to Queen Victoria. She had the 186-carat gem recut to a 108.93-carat oval brilliant. The Koh-i-Noor, set in the Queen Mother's crown, is on view in the Tower of London.

RIGHT: The Koh-i-Noor diamond crown worn by Queen Alexandra at her coronation, 1902.

the De Beer brothers discovered diamonds on their farm, and the Kimberley mine began production in what proved to be the richest deposit ever found. Unmechanized mining produced a huge crater known as "The Big Hole." Bucket load by bucket load, 25 million tons of earth were excavated to recover about three tons of diamonds.

In 1888, De Beers Consolidated Mines, Ltd. was formed by Cecil Rhodes, who consolidated the unwieldy claims at Kimberley, thus establishing the De Beers monopoly. De Beers maintained this virtual monopoly for more than one hundred years, distributing up to 85 percent of the wholesale supply of gem rough through its Central Selling Organization (CSO). In the second half of the twentieth century, the discovery of major diamond deposits in Russia, Africa, Australia, and Canada—not exclusively controlled by De Beers—diversified the marketplace.

With these nineteenth- and twentieth-century discoveries, diamonds have reached a popularity never before enjoyed. Ownership of the gem that until the fifteenth century had been reserved for royalty has become a realistic goal for the average person.

Legends and Lore

Hindus divided diamonds into four categories, mainly based on color, that could be owned by men of the four major castes; each category brought special good to its possessor: the Brahmin (priests), power, friendship, and wealth; Kshatriya (landowners and warriors), everlasting youth; Vaisya (merchant class), success; Sudra (laborers), good fortune.

A diamond is protective of its owner, according to a mid-fifth-century Sanskrit manuscript. It wards off serpents, fire, poison, sickness, thieves, flood, and evil spirits. Another Hindu belief was that a flawed stone has quite opposite effects; it could deprive even the god Indra of his highest heaven and could cause lameness, jaundice, pleurisy, and even leprosy.

The virtues of a diamond are legion. The stone provided fortitude, courage, and victory in battle, and it stood for constancy, purity, and enhanced love between husband and wife. In a fourteenth-century lapidary attributed to Sir John Mandeville, a diamond is claimed to lose its magical power because of the sin of its wearer. Two centuries later, the Italian mathematician and physician Girolamo Cardano was cautious: Whereas a diamond might make its owner fearless, fear and prudence might contribute more to well-being and survival, he noted. In addition, he stated that a diamond's brilliance irritates the mind just as the sun irritates the eye.

For hundreds of years, the belief that diamonds had gender persisted. Theophrastus (ca. 372–287 BCE) divided each species into male (the dark-colored stones) and female (the light-colored stones). As late as 1566, François La Rue, a French physician, described two diamonds that produced offspring.

During the Middle Ages, physicians debated whether diamond was a potent poison or an antidote to poison. The poison theory was refuted by Portuguese Garcia de Orta, physician to the viceroy of Goa; in describing the Indian mines in 1565, he noted that slaves working in the mines swallowed diamonds in order to steal them and showed no ill effects.

Occurrences

Diamonds require pressures in excess of 50,000 atmospheres to grow. The pressure corresponds to a depth of more than 90 miles, which is within Earth's upper mantle. Kimberlite, an unusual volcanic rock, is the mantle-sourced host for most diamonds. No kimberlite "volcanos" have erupted in over 30 million years, but if one did, the event would be devastating. The kimberlite magma ascends relatively rapidly from Earth's mantle, perhaps 30 miles per hour, but the gaseous eruption velocity exceeds the speed of sound, 740 miles per hour, like a high explosive. Kimberlite eruptions form carrot-shaped vertical volcanic vents, called "pipes," near the surface.

Around Kimberley, South Africa, kimberlite weathered into a surface layer of "yellow ground" and a lower layer of "blue ground" that contain intact diamonds. Further erosion carried diamonds into streams, rivers, and eventually beaches where, due to high density, diamonds became concentrated in placers. Placer deposits in India and South America were discovered in the search for placer gold.

In the most recent tabulation (2013), the Russian Federation was the world's largest producer of diamonds, by weight, but Botswana was the richest producer by value, 26 percent of the $14 million total. In fact, Africa is the diamond continent with 54 percent of the world's production by weight and 61 percent by value. Botswana is the second-largest diamond producer by volume, followed by the Democratic Republic of Congo, Zimbabwe, and Angola. Namibia is known for its productive beach placer deposits of diamonds—90 to 95 percent are of gem quality—so its mere 1.3 percent of world carats is nearly 10 percent of the value.

ABOVE: The 14.11-ct. emerald-cut Armstrong Diamond.

OPPOSITE: The Golden Maharaja, a 65.60-ct. diamond, was owned by one of the world's richest maharajas when it was shown at the 1937 Paris Exposition and at the 1939 New York World's Fair. The gem was featured in the Morgan Hall from 1976 to 1990 as an anonymous loan.

South Africa, once first in production, has dropped to eighth place in rough-diamond output. The giant Premier Mine has yielded many of the finest and most famous stones. Of the twenty largest diamonds found so far, ten were mined in South Africa. Other important African sources are Sierra Leone, Lesotho, Guinea, Tanzania, and Ghana.

Russia became a major diamond producer after scientific exploration in the 1950s and '60s located diamond pipes in ancient continental fragments in Siberia. The sources are primary, with the rich pipes including the well-known Mir, Udachnaya, and Yubileynaya in central Siberia. Active exploration continues in Siberia and northwestern Russia.

Canada is the diamond-discovery phenomenon of most recent time. Diamonds had long been thought to occur in Canada because of its geology and diamonds having been found in glacially transported deposits of the upper Midwestern United States. Prospecting in the 1980s led to the discovery of pipes that have been developed into mines, including Ekati, Diavik, Snap Lake, and Victor. These mines produce a large proportion of gem-quality crystals.

India's alluvial deposits were the first ever exploited, and many great historic diamonds were found there. The name Golconda, the famed ancient source, is synonymous with mine wealth. Indian production declined in the eighteenth century, although there is still small-scale mining and exploration. Brazil is an old but presently minor diamond supplier.

CONFLICT DIAMONDS

A modern look at diamonds as gems must include the role of diamonds in civil conflicts in Africa. In the 1990s, warring armies, particularly in Sierra Leone, Liberia, and Angola, enslaved workers to mine for diamonds that could then be traded for weapons. Trade in these stones besmirched the entire industry, so negotiations among interested parties and governments led to the so-called Kimberley Process Certification Scheme in 2002, whereby rough diamonds would be certified at mines and through trading as being "conflict free." This process continues and is reported to be effective for the vast majority of diamonds entering the marketplace.

Evaluation

A diamond's value depends on "the four *c*'s": carat weight, color, clarity, and cut.

Carat Weight. A 2-carat diamond costs more than twice as much as two 1-carat diamonds of the same quality and substantially more than four times four ½-carat diamonds. This fact is a manifestation of the greater rarity, and value, of larger diamonds.

Color. Absolutely colorless "white" stones are graded on the Gemological Institute of America's color grading scale as D. The letter grades run from D, E, and F on to S through X, at the bottom, for noticeably yellowish stones. To judge a diamond's color, look at it through its side against a white background. The gem should not be examined in direct sunlight, because fluorescence can mask a diamond's true color. The most valued natural fancy diamonds are bright red, but all vivid colors command high prices. Diamonds that are colored by irradiation and heating—green, yellow, golden brown, blue, purple, and red—require a disclosure.

Clarity. Most diamonds contain natural inclusions; by international agreement, a diamond is regarded as flawless if no inclusions are visible to the trained eye through a lens or loupe with tenfold magnification. Increasingly visible inclusions diminish the quality and grade of a stone. Vaporizing inclusions with a laser and filling of cleavages and fractures

to reduce their visibility is a valid enhancement process, but such practices must be disclosed.

Cut. Cutting brings out the full brilliance and fire of a diamond. The diamond should be faceted so that the maximum amount of light entering the stone reflects from the back facets and emerges back through the top. Poorly proportioned stones lose a lot of light through the back facets. The round brilliant cut was developed for diamonds and is the most popular cut, because it displays brilliance well. The oval, pear, and marquise cuts, which are modified brilliant cuts, appear larger than a round brilliant of the same weight but do not attain the same level of brilliance. The emerald cut, also called the step or trap cut, also yields reduced brilliance so it is often used for large flawless diamonds that would be blindingly brilliant if cut round.

SYNTHETIC DIAMONDS AND DIAMOND SIMULANT

Gem-quality synthetic diamonds greater than a carat in size were first made in 1970 but only recently have been grown in commercial quantities. A new process for growing diamonds, called chemical vapor deposition, was developed in the 1990s and made them much less expensive to synthesize. So, whereas synthetics had been more expensive than natural diamonds, that is no longer true. Cubic zirconia, synthetic zirconium oxide, and synthetic moissanite, silicon carbide, are the principal diamond simulants and are produced in both colored and colorless varieties.

OPPOSITE: The Lounsbery diamond necklace, designed by American philanthropist Richard Lounsbery (1882–1967) for his wife, Vera, and executed by Cartier of Paris. The necklace contains more than 100 diamonds, fashioned in rose, brilliant, pendeloque, and modified single cuts.

CORUNDUM

Sapphire and ruby are the gem varieties of the mineral corundum, but few people anticipate the myriad of colors displayed in these gemstones. Ruby is red, as everyone knows, and sapphire comes in all colors except red: pink, orange, yellow, brown, green, blue, purple, violet, black, and colorless. (Sapphire colors other than blue are termed "fancy" colors.) These are all represented in the Museum as in no other public display—our suite of large sapphires, some exceeding 100 carats, is famous. The huge 563-carat Star of India sapphire is one of the Morgan gifts. Its name suggests a story; one might speculate that, after being mined in Sri Lanka in the sixteenth century, it circulated among the treasuries of Indian potentates. If we only knew the sights this stone may have seen! However, George F. Kunz recorded only this enigmatic statement: "[It] has a more or less indefinite historic record of some three centuries and many wanderings." How the gem came into Kunz's hands is unrecorded, but rumor has it that a royal owner needed cash without publicity. An alternate, but doubtful, story is that Kunz had the stone fashioned in New York City in 1900—so much for romance! No matter, the Star of India is magnificent.

CORUNDUM DATA

Aluminum oxide: Al_2O_3

Crystal symmetry: Trigonal

Cleavage: Poor in one direction

Hardness: 9

Specific gravity: 3.96–4.05

R.I.: 1.76–1.78 (moderate)

Dispersion: Moderate (0.018)

OPPOSITE: The Star of India, the most famous gem in the Museum, is the largest gem-quality blue star sapphire in the world. The 563.35-ct. stone is nearly flawless and exhibits a perfect star. The almost spherical stone also displays a good star on its back side.

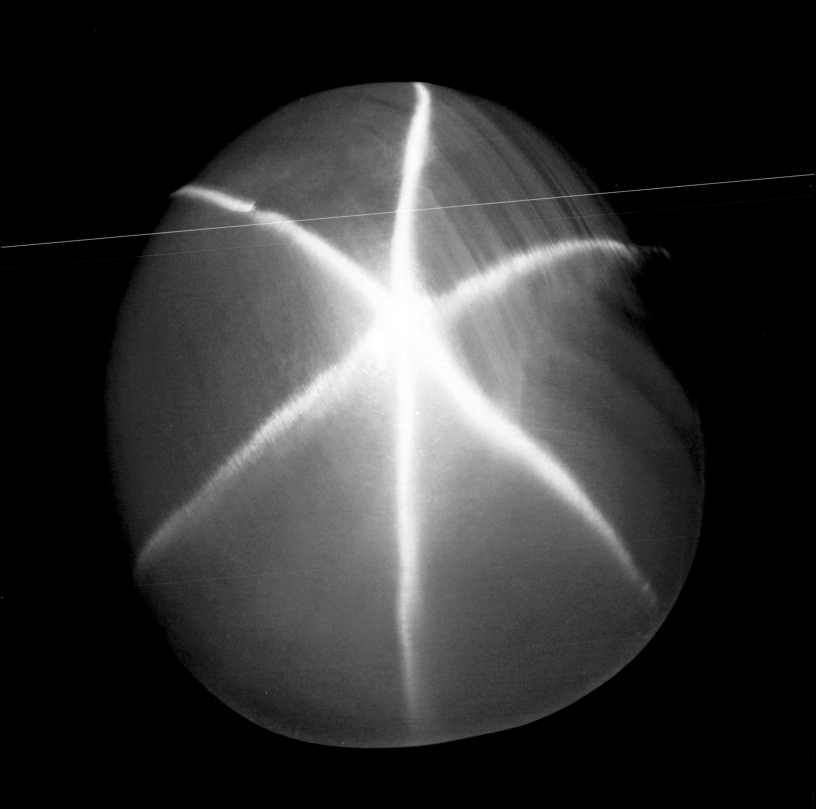

Properties

Spectacular colors, combined with great durability and reasonable abundance, have made sapphires and rubies important gemstones for centuries. Corundum has a tightly bonded structure that results in high density and great hardness, second only to diamond. The mineral is very stable chemically and essentially has no cleavage.

The great range of colors in sapphire and ruby is produced when aluminum is replaced in corundum by transition elements. Corundum gems are pleochroic, with colors being more intense when the crystal is viewed down the trigonal axis. Some stones manifest color-change

ABOVE: Small Sri Lanka sapphires, ranging from 2.00 to 16.90 cts.

pleochroism from purple to blue or, like alexandrite, from blue green to red.

Titanium is incorporated into the crystal structure of corundum at high temperatures and, during slow geologic cooling, it crystallizes separately as fine, silky fibrous inclusions of rutile (TiO_2). The result is chatoyancy that is multiplied by corundum's trigonal symmetry into six- and occasionally twelve-rayed star sapphires or rubies. These inclusions also make the gemstone translucent. Heat treatment "dissolves" the silk back into the corundum structure

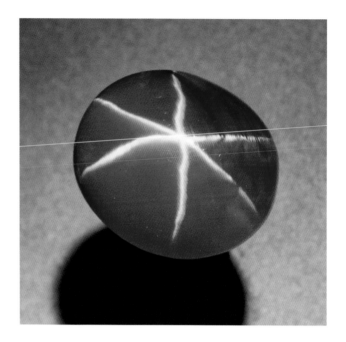

and generally results in a transparent, more intensely colored stone. We could heat the Star of India and create a fabulous blue cabochon—perish the thought!

Corundum crystals are typically hexagonal prisms that are somewhat tapered, often tabular, and frequently marked with triangular striations on the ends.

ABOVE: The Midnight Star ruby, 116.75 cts., is notable for its deep purple-violet color. The stone was found in Sri Lanka.

ABOVE RIGHT: The DeLong Star Ruby, 100.3 cts. One of the great star rubies, it was discovered in Burma (Myanmar) during the 1930s and donated in 1938 by Mrs. George Bowen DeLong. The ruby was stolen from the Museum, along with more than 20 other gems, in the "great jewel robbery" of October 1964. Ten months of intricate negotiations involving underworld figures and a ransom of $25,000 followed before the famous ruby was returned.

GEMSTONE CORUNDUM VARIETIES, COLORS, AND SOURCES OF COLOR

Ruby: **Intense red**—chromium

Sapphire: **Blue**—iron + titanium

Other than blue (called fancy sapphires):

Padparadscha: Orange–chromium + color centers (defects)

Alexandrite-like: Vanadium

Yellow: Ferric iron or color centers (defects)

Green: Combined yellow and blue mechanisms

Colorless: Pure, no substitutions

ABOVE: A ruby crystal, 4 cm (1½ in) long, in white marble, from Jegdalek, Afghanistan.

OPPOSITE: Two intergrown waterworn sapphire crystals, 5 cm (2 in) long, from Sri Lanka.

Historic Notes

Precious stones were grouped by color in antiquity. *Carbunculus* was the Latin term that Pliny the Elder used for transparent red stones. Before 1800, red spinel, red garnet, and ruby were all termed ruby, deriving from the Latin *rubeus*, meaning "red." A number of famous "rubies" have turned out to be spinels; the Black Prince's Ruby, the Timur Ruby, and the Côte de Bretagne Ruby are examples. The Catherine the Great Ruby, considered for years to be the largest ruby in Europe, was identified as tourmaline. *Sapphirus* is Latin for "blue," and until the Middle Ages, this name was used for lapis lazuli.

The history of sapphire dates back to the seventh century BCE, when it was used by ancient Etruscans. Following centuries saw the gemstone used in Greece, Egypt, and Rome. Rome obtained rubies from what is now Sri Lanka and from India, where ruby was the most valued of gems and called "king" and "leader" of precious stones. However, preserved Indian literature on corundum gems does not appear until sometime between 250 and 500 CE.

Marco Polo's travels took him to the "Island of Serendib" (Sri Lanka), and his thirteenth-century *Book of the Marvels of the World* gives high praise to both stones. He tells the story of a Sinhalese king, a ruby, and the Chinese emperor Kublai Khan. The ruby was huge—four inches—and Kublai Khan offered an entire city in exchange for it. The king refused, saying that he would not give up his prize for all the treasures in the world. Nothing more is known of this stone, and its reported size leads one to speculate whether it really was a ruby—or just a story.

Sapphire was a favorite stone for rings, brooches, and crowns of the medieval kings in Europe, and, beginning in the eleventh century, it also became the preferred stone for ecclesiastic rings. By the time of the Renaissance, both ruby and

LEFT: A ca. 1650 watercolor depicting Shah Jahan in Durbar, the Mogul court in India, holding a ruby; an attendant to the left holds a tray of jewels.

sapphire had found favor with the wealthy; indeed, only the wealthy could afford them. Florentine sculptor and goldsmith Benvenuto Cellini, writing in 1560, stated that the price of ruby was eight times that of diamond. And ruby is still generally the most valuable gemstone.

ABOVE: A portrait of Johannes the Steadfast, Elector of Saxony, by Lucas Cranach the Elder, from 1526; note the sapphire ring on his thumb.

Legends and Lore

The powers that have been ascribed to the ruby over the centuries are innumerable. Early Burmese thought the stone would bestow invulnerability when it was actually inserted into the owner's flesh. During the Middle Ages, a ruby was believed to have an inner fire that could not be concealed. Italian mineralogist Camillus Leonardus scorned those who denied the magical powers of precious stones and in the sixteenth century wrote that the ruby would preserve its owner's health, remove evil thoughts, control amorous desires, dissipate pestilential vapors, and reconcile disputes. Another belief was that a ruby could warn its owner of impending misfortune or calamity by becoming dull and dark. Catherine of Aragon (1485–1536), the first wife of Henry VIII, is said to have foretold her own downfall by perceiving the darkening of her ruby.

Sapphire's powers were equally sweeping. In a first- or second-century treatise ascribed to Damigeron, the sapphire protected kings from harm and envy. Its powers included the capacities of banishing fraud and preventing terror according to Marbode in the eleventh century. Two hundred years later, a French manuscript reported that the stone had the power of preventing poverty, while another lapidary of the same period stated that sapphire makes a stupid man wise and an irritable man good-tempered.

The star sapphire has been called "the stone of destiny"; its three crossed lines represented faith, hope, and destiny. Still another legend refers to these gems as sparks from the Star of Bethlehem. To the Germans, it was *Siegstein*, meaning "victory stone," according to Anselmus de Boodt, writing in 1609.

The Hindus, Burmese, and Ceylonese (Sinhalese) recognized a relationship between sapphires and rubies long before the Europeans did. To them, the colorless sapphire was an unripe ruby; if buried in the ground, it would mature and turn red—a belief documented in a sixteenth-century manuscript by Garcia de Orta. Flawed stones were considered overripe.

Occurrences

The primary sources are of two major types. First, high-temperature metamorphism of claystone and dirty limestone can form all corundum gemstones. Second, sapphires are found in some quartz-free igneous rocks. After weathering from primary sources, corundum gemstones concentrate in placer gravels—the most important commercial deposits.

The earliest sources for both ruby and sapphire are placers in Sri Lanka. Mining near Ratnapura (Sinhalese for "City of Gems") is referred to in the *Mahavamsa*, the great historical chronicle of Sri Lanka from the fifth or sixth century BCE. The rubies found here are paler than Burmese

stones. This is the original source of the rare lotus-colored (pink-orange) padparadscha sapphire and of the finest star sapphires. The blue sapphires are usually light in color.

Today the world's major sources of rubies are primary and alluvial deposits in East Africa, principally Mozambique, Tanzania, Madagascar, and Kenya. The quality is generally not the best, with a mixture of abundant flaws or turbidity and off-colors, but fine ruby gemstones have been produced particularly from Mozambique in recent years. Thailand near the Cambodian border used to be a major source but has not been so since the 1990s.

The world's finest rubies come from the Mogok Stone Tract in northern Myanmar (Burma)—the finest deep red (pigeon-blood) rubies, occasionally called "Burma rubies" regardless of source, and also dark or pale rubies. The deposits also yield sapphires and many other gems. Mogok

ABOVE LEFT: A large Burma ruby cabochon, 47 cts., a round 1.38-ct. ruby from North Carolina, and a 1.87-ct. ruby from Tanzania.

ABOVE RIGHT: This superb 100-ct. Sri Lankan stone is the largest and finest padparadscha sapphire on public display. The varietal name originates from a Sanskrit word for the orangey-pink color of the revered lotus.

mines have been producing rubies since about 600 CE and the present output is limited. Rubies with bluish zoning were discovered in Mongshu, Myanmar, around 1991. These stones respond very well to heat treatment and are produced in greater abundance than rubies from Mogok. For the United States, importation of rubies (and jadeite jade) from Myanmar was made illegal (U.S. JADE Act of 2008) in response to alleged human-rights abuses. Change from a military to civilian government may lead to the repeal of this law but has not done so by the editing of this book.

In recent years Australia and Madagascar have alternated as the world's leading producer of sapphires. In 1997, about 42 percent of the world's output came from New South Wales and Queensland. The sapphires are alluvial, weathered from basalt. The blue stones are generally dark and have an inky appearance. Major discoveries in Madagascar in the late 1990s have led to its gaining about 18 percent of the market as compared to 21 percent for Australia, 16 percent for Sri Lanka, and 14 percent from Kenya (compilation of 2005).

The major American source of sapphire is Yogo Gulch in Montana, which was discovered in 1895 and has been worked intermittently. The stones are small and waferlike and clear blue to violet. Sapphire gravels occur along the Missouri River near Helena, Montana, but even in good times production was only about 1 percent of the world market.

Other important sources of ruby and sapphire are Brazil, Vietnam, Nepal, Pakistan, India, and Cambodia.

ABOVE: Typical Montana "Yogo" rough and cut sapphires in blue and violet. The thin tabular gems range from .75 to 2.25 cts. in weight.

Evaluation

Color quality is most important to rubies and sapphires. Rubies with intense and uniform red to slightly purplish red—the so-called pigeon-blood color—are the most valuable. Medium-deep cornflower-blue sapphires are the most highly prized, and evenness of color is extremely significant. Orange and alexandrite-like sapphires of good quality command very high prices.

Flaws diminish the value of both gems; however, a fine-colored ruby is of high quality even if a minor flaw is present. To improve color and transparency, about 90 percent of new-mined sapphires are heat-treated with permanent results.

Star rubies and sapphires must be at least translucent to be of gem quality (with the exception of black-star sapphires). The star must have well-defined, sharp, straight rays that intersect at the center of the stone.

Large gem-quality rubies are rarer than large diamonds, emeralds, or sapphires. Thus the value of ruby, even more than of other gems, increases with weight.

Many other gems look like rubies and sapphires and are readily confused with and substituted for them. Synthetic corundums have been on the market since 1902 and are widely used in less-expensive jewelry.

GEMSTONES CONFUSED WITH GEM CORUNDUM AND THEIR TRADE NAMES

Ruby: Spinel (Balas ruby), pyrope garnet (Cape and Arizona ruby), red tourmaline (Siberian ruby), and pink topaz (Brazilian ruby)

Blue sapphire: Benitoite, cordierite, kyanite, spinel and synthetic spinel, and tanzanite

Green sapphire: Zircon

Yellow sapphire: Chrysoberyl

OPPOSITE: A fabulous assortment of Sri Lanka sapphires, ranging in weight from 18.19 cts. to 188 cts. The large 112-ct. yellow stone (upper right) was a gift in honor of Charles Zucker.

BERYL

Beryl is a mineral with several colorful gemstone varieties—aquamarine, morganite, heliodor—but the preeminent one is emerald. A popular misconception is that the Mogul, Ottoman, and Persian emerald treasures come from Asian deposits. In fact, the vast majority of these fabulous gems are from Colombian mines, Spanish loot from the New World. Spain needed money and found buyers among the Ottomans in Turkey and the Mogul nobility of India. Later conquests of the Moguls brought the treasures back to Persia. The Museum's Schettler Emerald (see page 45) is an example; it was cut in India during the period of Mogul domination and probably worn as headgear or a sleeve ornament by an Indian prince.

Its uncut counterpart is the Patricia Emerald. The largest gem-quality crystal on record from the famous Colombian Chivor Mine was discovered in 1920 and sold the following year for $60,000. The owner of the mine named the crystal after his daughter—and the name stuck. Only a few large fine crystals have been preserved in museums and bank vaults—emeralds are so valuable as gems that the crystals rarely escape being cut.

BERYL DATA

Beryllium aluminum silicate: $Be_3Al_2Si_6O_{18}$

Crystal symmetry: Hexagonal

Cleavage: None

Hardness: 7.5–8

Specific gravity: 2.63–2.91

R.I.: 1.566–1.602 (low)

Dispersion: Low

OPPOSITE: The Patricia Emerald is a twelve-sided crystal, 6.6 cm (2⅗ in) long, from the Chivor Mine, Colombia, and named for the mine owner's daughter. It weighs 126 grams (4.4 oz) and is famous for its crystal perfection and superb color as well as its size.

Properties

Beautiful color distinguishes the beryl varieties as gemstones. Beryl is a hard mineral but has only moderate brilliance and little fire. Elemental substitutions for aluminum in the crystal structure are the most common sources of color. However, there is a cavity along the six-fold axis in the structure that frequently accepts a chromophoric metal like iron, as in the case of aquamarine and some green beryl. Color-zoned crystals are possible, the most interesting being bicolor morganite-aquamarine.

Crystals form distinctive hexagonal prisms that, when from pegmatites, can be among the largest gemstone crystals. Inclusions vary with occurrence and variety. Emeralds are typically heavily flawed with cracks and inclusions of fluids and minerals from rocks in which they grew; these are called *jardin* (French for "garden"), the inclusion patterns resembling leaves and branches. The other beryl varieties usually have greater clarity, most commonly containing small parallel fluid-filled tubes that have the appearance of rain when a crystal is strongly illuminated. If these inclusions are fibrous, they can yield chatoyancy in a polished gem.

ABOVE RIGHT: Aquamarine, morganite, and heliodor from various localities, ranging in weight from the 11.38-ct. emerald-cut golden heliodor to the 390.25-ct. aquamarine.

GEMSTONE BERYL VARIETIES, COLORS, AND SOURCES OF COLOR

Emerald: Intense green or bluish green—chromium and/or vanadium

Aquamarine: Greenish blue or light blue—iron

Morganite: Pink, peach, or purple pink—manganese

Heliodor or Golden beryl: Golden yellow to golden green—iron + color center (defect)

Green beryl: Light green or too pale to qualify as emerald—iron, chromium, and/or vanadium

Red beryl or Bixbite: Raspberry red—manganese

Colorless beryl or Goshenite: Pure beryl, sometimes with cesium

Historic Notes

Of the gem beryls, the emerald has the longest history. The term *emerald* derives from the Greek *smaragdos*, for which there are conflicting meanings and antecedents. The earliest emeralds, extracted by Egyptians, date from the Ptolemaic era (323–30 BCE); however, tools dating to Rameses II (ca. 1300–1213 BCE) or even earlier have been found in tunnels at Sikait and Zabara, the location of ancient emerald mines in Egypt. These mines were probably the West's principal source for emeralds until the sixteenth century, although another source was known to the Celts

ABOVE: The Schettler Emerald weighs 87.62 cts. and its longest dimension is 3.5 cm (1⅜ in). It is engraved on both sides in a flower-and-leaf pattern and is probably from Muzo, Colombia.

ABOVE RIGHT: A platinum pendant set with diamonds and containing an emerald, 1.6 cm (⅝ in) long.

and Romans—Habachtal, south of present-day Salzburg in Austria. (The Archbishop of Salzburg had the deposits worked during the Middle Ages.) However, scientific research on some Gallo-Roman emeralds has suggested the source as Swat-Mingora in present-day Pakistan, rather than the two others.

Prior to the Spanish invasions in South America, emeralds from Colombia were traded and prized from Mexico to Chile. Almost from the time of their arrival, the conquistadors observed native rulers wearing them; in 1533 alone, Pizarro sent back four chests of emeralds to Spain. Finding no source in Peru, the Spaniards looked farther north and in 1537 discovered Chivor, in what is now Colombia. Following a skirmish with the Muzo Indians, from which the Spanish were forced to retreat, a soldier found an emerald and delivered it to his commander. The Spanish returned in force, defeated the Indians, and took

over the Muzo source. The Colombian gems were larger and of finer quality than any seen in Europe and Asia before the Conquest. They totally supplanted the other emeralds because of their superior quality.

The histories of aquamarine, heliodor, and morganite are more recent. The first documented use of aquamarine is by the Greeks between 480 and 300 BCE. The gem has been very popular since the seventeenth century. Heliodor (golden beryl) derives from two Greek words meaning "sun" and "gift"; this gemstone has also been known since antiquity. It has rarely been used in jewelry, for its color is not outstanding among other yellow gems. Morganite is the latest

gem to be recognized in the beryl group. First mined in Madagascar in 1902, the gem was named after J. P. Morgan by George F. Kunz.

ABOVE LEFT: Heliodor crystal 7.5 cm (3 in) long from Siberia.

ABOVE RIGHT: This is a 5.28 kg (11.6 lb) fragment from the largest aquamarine crystal ever found. The hexagonal prism, weighing 110.5 kg (243 lb) and measuring 48.3 x 40.6 cm (19 x 16 in), was discovered near Marambaia, Brazil, in 1910, and was so clear that, looking down the long axis, one could read a newspaper through it. The gemstone was cut in Idar-Oberstein, Germany, and yielded about 200,000 cts. of gems. A 47.39-ct. stone from Siberia is shown for scale.

owner eloquent and persuasive, and brings him joy. On the other hand, the emerald was considered an enemy of sexual passion, and, in the thirteenth century, Albertus Magnus wrote that when King Bela of Hungary embraced his wife, his magnificent emerald broke into three pieces.

The word *aquamarine* derives from two Latin words meaning "water" and "sea." Aquamarine amulets were thought to render sailors fearless and protect them from adversities at sea, especially if the stone were engraved with Poseidon on a chariot. The stone was a symbol of happiness and eternal youth, and according to Christian symbolism, it signified moderation and control of the passions to its owners. In medieval Europe, heliodor was believed to cure laziness.

Legends and Lore

To the Romans, the emerald symbolized the reproductive forces of nature and was dedicated to Venus; to the early Christians, it represented resurrection. In the fourth century BCE, Theophrastus noted its power to rest and relieve the eyes. Far later, Anselmus de Boodt (1609) recommended emerald as the most powerful amulet to prevent epilepsy, stop bleeding, cure dysentery and fever, and avert panic.

In addition, an emerald was thought to give its owner the ability to foretell the future. According to Marbode (eleventh century), emerald improves memory, makes its

ABOVE: The official trapping of a vizier of Morocco (ca. 1750). Aquamarines set in gold and surrounded by small diamonds, rubies, sapphires, and red garnets.

51

Occurrences

Emeralds are most frequently found in metamorphosed shales, particularly mica schists—the reason for some emeralds containing mica inclusions. In Colombia, emerald deposits occur in calcite veins in black shale (Muzo) and in quartz veins in limestone (Chivor). Classification of emerald deposits is difficult because the conditions that combine the geochemically dissimilar elements beryllium and chromium are not systematic but results of a combination of tectonic activity and thermally driven fluids passing through dissimilar rocks. Aquamarine, morganite, and heliodor are found as well-formed crystals in pegmatites. Beryls are not sufficiently dense to concentrate in placers and are normally mined from the primary source or its weathered equivalent.

Colombia is the world's largest emerald producer with numerous mines in operation. Muzo (also Coscuez, Maripí, La Pava, Peñas Blancas, and Yacopí) and Chivor (also Gachalá and Macanal) are the two principal mining areas. Muzo yields the world's finest emeralds. Mining operations have continued there since the Spanish Conquest almost without interruption. In general, Chivor emeralds are less flawed but do not have as velvety an appearance as those from Muzo.

Mining of emeralds in Russia began shortly after a gem crystal was discovered in 1830 by a peasant in the Ural Mountains northwest of Yekaterinburg (Sverdlovsk). He took his find to the lapidary factory in Yekaterinburg, where geologist Yakov Kokovin recognized it as an emerald. Some years later, Kokovin's office was searched, and a large emerald crystal was found. Kokovin was sent to prison,

ABOVE: Well-formed emerald crystal 6.5 cm (2½ in) long, from Takowaja, Ural Mountains, Russia.

where he committed suicide. The 2,226-gram (5-lb.) crystal now resides in the Mineralogical Museum of the Academy of Sciences in Moscow. The Izumrudnye Kopi emerald district emeralds vary from fine deep green (but heavily included and flawed) to yellowish green (and less included). Production from the Malyshev mine is variable but further development has been proposed in recent years.

Zambia emerged as a source of emeralds in 1977 and is now the number two producer worldwide, with some very fine gems. Brazil has also become a significant producer of emeralds from Itabira/Nova Era (Minas Gerais), Carnaíba-Socotó (Bahia), and Santa Terezinha (Goias). Small fine stones have come from the Sandawana mine in Zimbabwe, and the Panjshir Valley of Afghanistan may be poised to become a significant producer. North Carolina is the most important source in the United States, but there is no steady supply. The best-known locality is around Hiddenite, where emeralds were discovered in 1880.

Brazil is the principal source of aquamarine. More than 80 percent of the country's aquamarine comes from an area around Teófilo Otoni in the western part of Minas Gerais. Santa Maria aqua is the name of fine-colored material from the Santa Maria de Itabira mine, Minas Gerais, but similar fine color from Africa, particularly Mozambique, has also been sold with that name. Thus, Santa Maria Africana is the preferred term for the later gems. Madagascar is known for the rich blue color of its aquamarine, which resembles sapphire. Somewhat pale aquamarine from Afghanistan and Pakistan became abundant in the 1980s, particularly during the Afghan War, and remains in plentiful supply. Aquamarine is also found in the Ural Mountains of Russia, in Transbaikalia, and in Siberia. Other occurrences of aquamarine include China; Vietnam; Malawi; Nigeria; and Zambia; and in the United States, in Maine, Idaho, and California.

RIGHT: An etched aquamarine crystal 8 cm (3⅛ in) tall, associated with white albite, from the Dusso area in Pakistan.

The major sources of morganite are San Diego County in California, Minas Gerais in Brazil, and Madagascar. Heliodor is found in Minas Gerais and Goiás, Brazil; the Ukraine in Russia; and also in Connecticut and Maine.

Evaluation

Emeralds with a rich, velvety green uniform color and a minimum of flaws are considered the finest quality. Skillful cutting can both minimize the visibility of inclusions and bring out the gem's best color. Emerald is often oiled or infused with resin to conceal or minimize cracks, and sometimes dye is added to improve the color—a fraudulent practice.

Synthetic emeralds have been produced since 1934. Their prices are higher than those of other synthetics and even some natural emeralds.

Intensity of color and clarity are the most essential considerations in evaluating aquamarine, morganite, and heliodor. Aquamarine should be bright sky blue or sapphire blue. Aquamarines with intense color are becoming very scarce, and their price has increased substantially. The bright sky-blue shade can be produced by heat-treating greenish-yellow, greenish, and even brownish beryls. The color change is permanent. Gem-quality aquamarine is generally free from inclusions. Morganite should be deep purple pink, but peach-colored gems are next best. Heliodor with a deep yellow to yellow-green color is desirable.

GEMSTONES CONFUSED WITH GEM BERYLS

Emerald: Demantoid and tsavorite garnets, Imperial jade, tourmaline, peridot, green zircon, and hiddenite

Aquamarine: Blue topaz, euclase, kyanite, apatite, sapphire, tourmaline, and zircon

Morganite: Kunzite, tourmaline, topaz, sapphire, spinel, and rhodolite garnet

OPPOSITE: A matrix specimen consisting of a perfect morganite crystal, 6.5 cm (2.5 in) across, with a gem-quality bicolor elbaite from San Diego County, California, next to a square-cut 278.25-ct. morganite from Minas Gerais in Brazil.

ABOVE: This delicate aquamarine necklace with pearls and diamonds dates from the early twentieth century.

ABOVE: A superb 58.79-ct. morganite from Madagascar.

OPPOSITE: This Chinese carving of a sitting goddess 10.5 cm (4⅛ in) high is the finest and largest morganite carving known.

CHRYSOBERYL & SPINEL

Chrysoberyl

Alexandrite and cat's eye are renowned for their eye-catching properties, but few people know that they are varieties of a mineral that is more common in its transparent yellow-green gemstone form. All three are chrysoberyl. Cat's eye chrysoberyl is *the* cat's eye; when spotlighted, the gem exhibits a band of light that opens and closes as the stone is turned. Alexandrite changes from red in incandescent light to green in daylight. It is *the* color-change gemstone. By comparison, ordinary chrysoberyl, though fine in its own right, seems a poor relation.

CHRYSOBERYL DATA

Beryllium aluminum oxide: $BeAl_2O_4$

Crystal symmetry: Orthorhombic, pseudohexagonal

Cleavage: Distinct in one direction

Hardness: 8.5

Specific gravity: 3.68–3.78

R.I.: 1.74–1.76 (moderate)

Dispersion: Moderate

OPPOSITE: An 85-ct. cat's eye from an unknown locality.

one direction. Yellow, brownish, and green cat's eyes are most common; alexandrite cat's eye is very rare.

Individual chrysoberyl crystals as rectangular prisms are rare. Instead, intergrowths (twins) called *trillings* or *sixlings* with near-hexagonal symmetry are more abundant.

Historic Notes

The term *chrysoberyl* derives from the Greek *chrysos*, referring to the stone's golden color, and the mineral beryl. Until 1789, when A. G. Werner, a famous German geologist, identified chrysoberyl as a mineral species, the stone was erroneously assumed to be a variety of beryl. Another term for precious cat's eye is *cymophane*.

Cat's eye has the longest history of the chrysoberyl varieties. Although known in Rome by the end of the first century, it had been treasured even earlier in South Asia, where chatoyant stones have always had admirers. The gem was forgotten in the West until the late nineteenth century, when the Duke of Connaught gave a cat's eye betrothal ring to Princess Louise Margaret of Prussia. The gem's popularity—and price—rose immediately; Ceylon (Sri Lanka) had difficulty keeping up with the demand. Cat's eye is currently a fashionable ring stone, particularly in Japan and China.

Properties

Chrysoberyl is among the most brilliant gemstones, only surpassed in hardness by diamond and corundum. The common variety is transparent yellowish green to greenish yellow and pale brown, the result of small amounts of iron replacing aluminum. Chromium is the coloring agent in green chrysoberyl and alexandrite. Alexandrite's pleochroism, which is the same as its color change, is evident when the gemstone is viewed from perpendicular directions. The cat's eye is caused by fine needle inclusions of rutile (TiO_2) in

ABOVE LEFT: A chrysoberyl trilling, consisting of three twinned crystals, from Espirito Santo, Brazil; it measures 8 cm (3⅛ in) across (front and side view).

LEFT: The Duke and Duchess of Connaught, ca. 1907. The Duchess wore a cat's eye betrothal ring.

Alexandrite was discovered in 1830 in an emerald mine near Yekaterinburg (Sverdlovsk) in the Russian Ural Mountains on the birthday of the then–heir apparent czar Alexander II, for whom the gem was named. The naming was doubly appropriate not only because of the discovery date but because the chameleon-like green and red colors were the same as those of the Russian Imperial Guard.

The third and common variety of chrysoberyl, the transparent greenish-yellow form, was found in Sri Lanka and Brazil. The Brazilians called it *crisolita* and named a city in its honor. Exported to Europe, the gem became popular and was used in eighteenth- and nineteenth-century Spanish and Portuguese jewelry. This variety of chrysoberyl was in great demand during the Victorian and Edwardian eras but is now overshadowed by alexandrite and cat's eye.

Legends and Lore

The natives of Sri Lanka and India believed that cat's eye protected its wearer from evil spirits because the stone was inhabited by a "spirit of potent good" (see Kunz 1938). According to Hindu lore, it preserved the owner's health and guarded him or her against poverty. In Eastern belief, the gem would endow foresight if pressed against the forehead at a point between the eyes. The movement of the

eye, when light or visual orientation is changed, is thought to have been the origin of this belief.

Alexandrite has been regarded as a stone of good omen in Russia and is the only gem accorded the role of a talisman as recently as the nineteenth century.

RIGHT: Chrysoberyls from Sri Lanka and Brazil, ranging in weight from 8.9 to 74.44 cts.

GEMSTONES CONFUSED WITH CHRYSOBERYL GEMS, AND SYNTHETICS AND IMITATIONS

Alexandrite: Synthetic sapphire and synthetic spinel are widely used as (usually poor) imitations. Synthetic alexandrite has had only marginal market success in the United States.

Cat's eye: Quartz and occasionally fine tourmaline cat's eye are commonly confused with chrysoberyl cat's eye. The unmodified name "cat's eye" applies only to chrysoberyl. Imitation cat's eye can be made by cutting synthetic or natural star sapphire in such a way as to display only one ray of the star. Cat's eye may also be imitated with triplets of the white fibrous mineral ulexite sandwiched between two pieces of yellow synthetic sapphire or glass.

Occurrences

Chrysoberyl crystallizes in and around pegmatites rich in beryllium, but deposits commonly form as alluvial concentrations from weathered pegmatites.

The major source of all chrysoberyl varieties is Minas Gerais in Brazil. Here, at Lavra de Hematita in 1987, the world's largest find of alexandrite produced, in less than three months, 50 kilograms of fine gems—some up to 30 carats.

Russian alexandrite deposits near Sverdlovsk and one discovered later near the Sanarka River in the southern Urals are apparently exhausted, even after post-Soviet processing of the mine waste. Sri Lanka produces cat's eye and alexandrite from its gem gravels. Generally, the alexandrites are larger than the Russian stones and have a more attractive daytime green color. However, the Russian alexandrite has a better color change and a finer red color under artificial light. Other occurrences of alexandrite and yellow chrysoberyl are in India, Madagascar, Tanzania, and Myanmar.

Evaluation

Only chrysoberyl with a distinct color change is alexandrite; either the red or green must be a good hue. Flaws diminish the value. Fine-quality stones exceeding five carats are both rare and expensive. Alexandrite cat's eye is one of the rarest and most costly of gems.

Cat's eye is the most valuable chatoyant gemstone. A rich honey yellow brings the highest price; the green stones follow in value. The chatoyant band must be sharp, narrow, and positioned in the center of the cabochon. In the finest stones, when the band is at right angles to the light, the half of the stone facing the light should appear milky; the other half should have a rich honey color. The eye should also open widely in oblique illumination and close sharply in direct illumination. Stones that approach transparency and exhibit a sharp eye command the highest price. Fine stones over 20 carats are rare and expensive. Of the common transparent variety of chrysoberyl, intensely colored yellowish green stones are the most popular and highly valued.

LEFT: An 8.9-ct. alexandrite from Sri Lanka, showing its color change— in daylight (green) and in incandescent light (red).

Spinel

For centuries, most gem spinels were thought to be rubies or sapphires—a reasonable assumption, because both spinel and corundum are found in the same deposits and have similar properties. A famous example is the Black Prince's Ruby. The stone's known history began in 1367, when it was taken from the treasury of the King of Grenada by the victorious Don Pedro of Castile. He presented the "ruby" to the Black Prince, son of Edward III, in payment for services at the Battle of Nájera in northern Spain. As part of the breakup of the crown jewels by the Commonwealth in the 1650s, the gem was sold (inventoried at four pounds sterling) and somehow returned to the monarchy during the Restoration. The 5-cm (2-in) irregular spinel resides at the center of the British Imperial Crown.

SPINEL DATA

Magnesium aluminum oxide: $MgAl_2O_4$
Crystal symmetry: Cubic
Cleavage: None
Hardness: 8
Specific gravity: 3.58–4.06
R.I.: 1.714–1.75 (moderate)
Dispersion: Moderate

Spinel is also the name of a mineral group of multiple oxides with the same crystal structure. Gahnite, with zinc replacing magnesium, is a blue spinel gemstone.

ABOVE: A 70.99-ct. spinel from Sri Lanka.

Properties

Like the corundum gemstones with which it is confused, spinel is known for its many colors and durability. Spinel is slightly softer than corundum because the magnesium-to-oxygen bonds in spinel are not quite as strong as the aluminum-to-oxygen bonds found in both minerals. Nevertheless, spinel's hardness, combined with no weak plane in its crystal structure, makes the gemstone very durable.

In another comparison to corundum, the presence of magnesium in addition to aluminum in spinel permits a wider range of chemical substitutions, transforming the pure colorless spinel into many possible colors but, surprisingly, not as many colors as corundum. Names have been applied

ABOVE: Ring with 9.5-ct. spinel from Myanmar. Other spinels, from Sri Lanka, range from 1.89 to 46.48 cts.

to many of the different varieties, but now they are referred to simply by their color—red spinel, green spinel, and so on. Synthetic spinels can be produced in an even greater range of colors by adding elements such as cobalt, manganese, and vanadium in amounts or combinations not found in nature.

Infrequently, spinels contain fine needles in three perpendicular directions that manifest four- and six-rayed stars in cabochon stones.

With cubic symmetry, spinel crystals are predominantly in the form of an octahedron. The term *spinel* may refer to this shape; the Latin *spina* means "thorn." Although spinels have probably been recognized by their octahedral crystals for many centuries, the distinction between ruby and spinel as different mineral species was not made until 1783 by French mineralogist Jean-Baptiste L. Romé de l'Isle.

Historic Notes

The earliest red spinel used as an ornament was found in a Buddhist tomb near Kabul, Afghanistan, and dates from about 100 BCE. Red spinel was also used by the Romans of the first century BCE. Blue spinels have been found in England dating from the Roman period (43–409 CE), and a ring set with a pale green octahedral spinel from the Eastern Roman Empire has been described.

Mining of spinel began in Badakshan, Afghanistan, sometime between 750 and 950. The locality was first described by the Arab geographer Istakhri in 951 and later by Marco Polo. Many of the historic spinels were probably mined here, and one interpretation is that the name for these spinels—Balas rubies—derived from the word for the former name of the region.

Of the spinels, the red stones have had the longest and most dramatic histories. Like the Black Prince's Ruby, the Timur Ruby's known history goes back to the fourteenth century. The stone bears six Persian inscriptions, the oldest identifying it as being in the possession of the Tartar conqueror Timur (Tamerlane) in 1398. Over the years, the stone was traded or plundered in India until the East India Company took possession in 1849 and presented it to Queen Victoria two years later. The stone is now in the private collection of Queen Elizabeth II.

Red spinels were used in Renaissance jewelry in Europe and became popular during the eighteenth century. Elegant pendants, earrings, and brooches set with spinels and diamonds survive from the former Russian and French crown jewels. In Moscow, a deep red 412.25-carat spinel surmounts the Great Imperial Crown commissioned for the coronation of Catherine the Great. The world's largest collection of spinels, including a record-holding 500-carat stone, is part of the former crown jewels of Iran.

Catherine the Great of Russia in her coronation robe and crown, in a 1778–79 painting by Danish painter Vigilius Eriksen.

Legends and Lore

The Hindus considered spinels to be rubies and divided them according to caste. The members of each of the four major castes should wear the appropriate stone in order to benefit from its virtues: the Brahmin priestly caste—true ruby; Kshatriya (knights and warriors)—rubicelle (yellow-to-orange tinge); Vaishya (landowners and merchants)—ruby spinel; Sudra (laborers and artisans)—Balas ruby (rose-red spinel).

In ancient and medieval times, when color had strong symbolism, red spinel and other red stones were considered cures for hemorrhages and all inflammatory diseases, as well as prescriptions to soothe inflamed emotions, eliminating anger and conflict. An Indian belief, reported by an Armenian writer of the seventeenth century, is that powdered spinel taken in a potion eliminated dark forebodings and brings happiness.

Occurrences

Gem spinels, like corundum, form in highly metamorphosed claystones and especially dirty limestones transformed into marbles. Most spinels are produced from alluvial concentrations from weathered primary sources.

The Mogok Stone Tract and Nanyaseik (Namya) in northern Myanmar are the source of the finest-quality spinels—rose, pink, orange, blue, and violet colors—which occur as waterworn pebbles and as perfect octahedra, particularly when found in the hosting marble. Luc Yen in Vietnam has produced similar-quality spinel as Mogok in the last decades. Sri Lanka's gem gravels are located in the southwest part of the island around Ratnapura ("City of Gems"). There the spinels, always waterworn, are generally blue, violet, or black (named "ceylonite" after the country's former name). Red, orange, and pink spinels are rare. Other occurrences are in the Ilakaka region of Madagascar, Thailand, Tanzania, and the Pamir Mountains of Tajikistan.

GEMSTONE SPINEL COLORS AND SOURCES OF COLORS

Red: Chromium (deep red—ruby spinel; rose red—"Balas ruby")

Purple red: Chromium + iron (almandine spinel)

Blue: Iron and/or cobalt (sapphire spinel and gahnospinel)

Green: Iron (chlorospinel)

Evaluation

Color, clarity, and weight are important considerations in appraising spinel. Red spinels are the most valuable, the most highly prized ones being orange red and intense red to purplish red. Because spinel is often flawless, its clarity is of great importance—much more important for spinel than for ruby, which is rarely found without inclusions. Stones weighing more than five carats are uncommon—the large ancient stones are no longer found.

Spinel possesses both beauty and durability, but confusion with the cheap synthetic version and insufficient supply prevents it from enjoying the popularity that it merits. In recent times, spinels have become rarer than rubies and sapphires in the marketplace, even though values of comparable quality spinels are not generally as high as for the corundum gems.

GEMSTONES CONFUSED WITH SPINEL AND SYNTHETIC SPINEL

Spinel can be mistaken for ruby, sapphire, pyrope garnet, amethyst, and zircon. Synthetic spinel, which is manufactured in large quantities, is inexpensive and is used to imitate ruby, sapphire, emerald, aquamarine, peridot, alexandrite, and diamond.

ABOVE: A 4.03-ct. cut spinel and an octahedral crystal 1 cm (⅜ in) across in marble. Both are from Mogok, Myanmar.

67

TOPAZ

Topaz is renowned for its capacity to form large gem-quality crystals. Pliny, in the first century, remarked: "Topazos [topaz] of all precious stones is the largest. In this, it excels all others." While in Minas Gerais, Brazil, in 1938, Allan Caplan, a New York mineral dealer, noticed with some excitement exceptionally large topaz crystals for sale. Upon learning the news, many United States museums began competing to obtain the "giants." The Smithsonian picked a topaz weighing 71 kilograms (156 lb); then the Cranbrook Institute got a slightly smaller one. Harvard, followed by the American Museum of Natural History, wanted one, but no more were available.

During his fourth trip to Brazil in 1940, Caplan learned of three prodigious specimens in transit to Rio de Janeiro. On the basis of photographs alone, he bought them and returned home to await the arrival of his prizes, which were revealed to weigh 271, 136, and 102 kilograms (596, 300, and 225 lb)!

Finally, after many months, the crates arrived. At the U.S. Customs Office, eagerly and anxiously, Caplan approached the largest crate and removed the top boards. To his horror, he saw a great broken surface staring up at him from the packing material! Bitterly disappointed, he had the crate resealed and all three specimens sent to the Museum for evaluation. There a group assembled for the grand opening. The moment arrived. The crate, now turned

TOPAZ DATA

Aluminum silicate fluoride hydroxide: $Al_2SiO_4(F,OH)_2$

Crystal symmetry: Orthorhombic

Cleavage: Perfect in one direction

Hardness: 8

Specific gravity: 3.5–3.6

R.I.: 1.606–1.644 (moderate)

Dispersion: Low

OPPOSITE: Imperial topaz crystal, 5.5 cm (2⅛ in) long from Ouro Preto, Minas Gerais, Brazil, and a 16.95-ct. cut stone from Sri Lanka.

topside down, was pried open, and the superbly terminated "top" of the great crystal came into view. All in attendance breathed a collective sigh of relief as they gazed at the largest fine topaz crystal in "captivity"—and ever since a treasured possession of the Museum.

Properties

Topaz is revered for its color, clarity, and hardness. Strong chemical bonding makes it relatively dense and the hardest silicate mineral. However, a weak plane in the crystal structure occupied by fluorine and hydroxyl is the source of one excellent cleavage. This cleavage is topaz's major failing as a gemstone and demands great care in cutting and handling.

A common misconception is that all topaz is yellow. In fact, pure topaz is colorless, and colors include blue, pale green, and the spectrum from yellow through the familiar sherry orange to pink and even the most rare red. Chromium substitutes for aluminum, producing red and some pink topaz, but most other colors are a result of minor atomic substitutions and defects in the crystal followed by radiation damage. Some of these colors are unstable and can fade; some browns fade totally in sunlight, and some sherry orange stones become pink upon heating. High-energy irradiation of colorless topaz followed by heat treatment at moderate temperature yields blue stones with stable color. Natural blue topaz appears to be created by an identical process, so there is no way to distinguish between natural and "created" blues. However, intense natural blue stones are not known, so it is a good bet that deep blue color in gem topaz is artificially produced.

Topaz crystals typically form prisms having a diamond-shaped cross section and a pyramidal top; the cleavage cuts straight across the prism.

ABOVE: The large faceted red topaz is an oval brilliant-cut gem of 70.40 cts. with an unusual natural deep red color. It is from either Brazil or Russia.

RIGHT: A page from the 1889 English translation of Jean-Baptiste Tavernier's *Travels in India* (1677), with an illustration of a large topaz in the center (fig. 6). Tavernier describes it as "the large topaz of the Great Mogul. I did not see him wear any other jewel during the time I remained in his court.... This topaz weighs 181⅛ ratis, or 157¼ carats. It was bought at Goa for...the sum of 181,000 rupees."

70

Historic Notes

How topaz came by its name is uncertain. *Topazos*, which means "to seek," was the Greek name of a fog-bound, hard-to-find island in the Red Sea, known in Europe as St. John's Island), but today as Zabargad (see page 81). Pliny the Elder considered topazos to be a green stone from that island.

However, since it is the green gemstone peridot—not topaz—that is found there, Pliny must have been describing peridot in this case. Another confusion occurs with the term *chrysolite*, identified both with peridot and yellow gems such as topaz. The Sanskrit word *tapaz* means "fire" and seems a more appropriate possibility for the derivation of the word *topaz*.

Topaz was used in ancient Egypt and Rome; the Romans obtained their topaz from Sri Lanka, an early and continuing source for the gemstone. In the seventeenth century, Jean-Baptiste Tavernier mentions both the stone and the location in accounts of his buying trips in the Orient.

During the Middle Ages in Europe, topaz was not particularly popular, although it was occasionally used in ecclesiastical or royal jewelry. But by the eighteenth century in Spain and France, the gem enjoyed increased popularity and, together with diamond, was set in many magnificent pieces of jewelry. Early in the next century, topaz and amethyst were the most stylish gems for earrings and necklaces in both France and England. Topaz continued to be one of the most popular gems during the Victorian era and later became a favorite stone of the Art Deco jewelers. Commonly regarded as the finest yellow stone, its popularity persists.

PLATE IV.

RUBIES AND TOPAZ.

Legends and Lore

During the Middle Ages, the topaz was thought to strengthen the mind and prevent mental disorders as well as sudden death. Marbode's eleventh-century poetic treatise recommends it as a cure for weak vision. The prescription called for immersing the gem in wine for three days and three nights, followed by application of the topaz to the afflicted eye. A topaz engraved with the figure of a falcon could help its bearer cultivate the goodwill of kings, princes, and magnates, according to a sixteenth-century lapidary entitled Speculum Lapidum by Camillo Leonardi (allegedly based on the Hebrew *The Book of Raziel*, but Kunz [1938] argues there is little similarity in the texts). Still later, topaz was recommended by Girolamo Cardano as a cure for madness, a means of increasing one's wisdom and prudence, and a coolant for both boiling water and excessive anger.

Topazes from various localities, ranging in weight from 17.16 cts. to 375 cts.

Occurrences

Topaz is found principally in gem pegmatites, where fluorine is often abundant. This volatile-rich environment stimulates the growth of large crystals. Weathering of these pegmatites releases topazes into streams and rivers; the gemstones concentrate in alluvial gravels.

Minas Gerais in Brazil is the world's largest producer of topaz—blue, colorless, and sherry colored. Topaz was discovered there first near Ouro Preto in 1735, the primary source of sherry-colored topaz. In the Ural Mountains, topaz is found northeast of Sverdlovsk in Mursinka and Alabashka and at Sanarka in the southern Urals. Another important site is Volodarsk-Volynskiy in Ukraine, the source of beautiful etched amber-colored crystals. Topaz is mined north of Katlang in Pakistan in veins of coarse-grained calcite and quartz in marble that yield topazes of many colors, including a rare pink and reddish brown. Colorless to pale amber topaz is mined from pegmatites in the Mogok Stone Tract, Myanmar. Other pegmatite sources are in Madagascar and Namibia.

ABOVE: The world's largest topaz crystal, 271 kg (596 lb), is from Minas Gerais, Brazil.

LEFT: Early nineteenth-century engraving of Ouro Preto (formerly Vila Rica), where topaz was first discovered in 1735.

OPPOSITE: Both the 47.55-ct. pink topaz and the 47.75-ct. Imperial topaz are from Minas Gerais in Brazil.

Evaluation

Two factors should be considered in evaluating topaz: color and clarity. The most valued color is a rare, almost unavailable, red. Imperial topaz, a sherry-colored stone, has always been the most popular. Both sherry-colored (brownish yellow, orangey yellow, and reddish brown) and pink topaz command high prices. Precious topaz is yellow topaz, a term commonly used to distinguish it from other topaz colors and from citrine. Light blue and pale yellow stones are of less value. The value diminishes significantly when the stone is flawed.

GEMSTONES CONFUSED WITH TOPAZ AND TRADE NAMES

Tourmaline, sapphire, chrysoberyl, and the rare danburite, andalusite, and apatite can be confused with topaz.

Yellow quartz, or citrine, which lacks topaz's velvety appearance, brilliance, and rich color, is occasionally sold as topaz—an unethical practice. Equally unethical is the substitution of "treated" blue topaz for the rarer and more expensive aquamarine.

Misleading trade names are frequently used in substituting other less valuable stones for precious topaz: they include Spanish, Saxon, and Bohemian topaz for citrine quartz; Smoky, Burnt, and Scotch topaz for smoky quartz; and Oriental topaz for yellow sapphire.

TOURMALINE

One spring morning in 1876, a young man walked briskly and confidently into Tiffany & Co. In the director's office, he unfolded a gem paper and placed what he later called a "drop of green light" on the desk. The "light" was a sparkling faceted green tourmaline from Maine, which spoke for itself. Both men admired its quality and beauty. Charles Tiffany bought it immediately, much to the delight of George F. Kunz. Within a year, the twenty-year-old gemologist had embarked on his illustrious career at the company. As the preeminent gem expert of his day, he championed lesser-known colored stones. Tourmalines were his favorites among American stones, and the timing of his visit was opportune. Several United States sites yielding commercially viable quantities of gem tourmaline had just been or were about to be discovered. Kunz collected tourmaline for many institutions, including the Museum—his Maine, Connecticut, and California tourmalines abound in the collection.

TOURMALINE DATA

Complex aluminous borosilicates

Elbaite: $Na(Li_{1.5}Al_{1.5})Al_6(Si_6O_{18})(BO_3)_3(OH)_3(OH,F)$

Dravite: $NaMg_3Al_6Si_6O_{18}(BO)_3(OH)_3(OH,F)$

Uvite: $(Ca,Na)(Mg,Fe^{2+})Al_5MgSi_6O_{18}(BO_3)_3(OH)_3(F,OH)$

Liddicoatite: $Ca(Li_2Al)Al_6(Si_6O_{18})(BO_3)_3(OH)_3(OH,F)$

Rossmanite: $\square(LiAl_2)Al_6(Si_6O_{18})(BO_3)_3(OH)_3(OH)$

Crystal symmetry: Trigonal

Cleavage: None

Hardness: 7–7.5

Specific gravity: 2.9–3.1

R.I.: 1.610–1.675 (moderate)

OPPOSITE: Bicolored tourmalines from Mesa Grande in California: three superb "pencil" crystals. The longest is 9.5 cm (3¾ in). The cabochon is 22.40 cts. and the cut stone is 30.50 cts.

Properties

Tourmaline is a large mineral group whose members display the broadest spectrum of gemstone colors. There are at least thirty-nine mineral species in the group (species have been added almost annually), but only five have found use as gems (or are found in gems—zonations lead to gems with multiple species). These are sufficiently durable (hard and free from cleavage) to be fine gemstones; however, many tourmaline crystals are internally strained and subject to cracking during faceting and setting in jewelry. Named for the Isle of Elba, where it was first found, elbaite is the tourmaline most often used in jewelry; dravite and uvite are less common and rarely of the appropriate quality but are also gemstones. Gem elbaite can be pink to red, blue, green, violet to red purple, yellow, orange, brown, black, and colorless. Bicolored crystals with one end green and the other pink are common. Crystals with a pink core and green rind are called "watermelon."

Color in gem tourmaline is primarily a result of substitutions of transition elements for other metals in the crystal structure. There are few useful generalities relating color to a specific chemical element, but pink is usually due to manganese, and green is attributed to ferrous iron, chromium, or vanadium. Color may be improved by heating and/or irradiation, but the changes are not always permanent.

Elbaite crystals are often recognizable by their prismatic form, typically elongated like pencils with cross sections that range between hexagonal and trigonal; crystals are often sufficiently perfect and clear to be natural gems. Color is created by changes in chemical composition in response to changing conditions during crystal growth, which yields both chemically and color-zoned crystals. Elbaite crystals grow fastest at their ends, changing color as they grow; layers of growth from core to rim produce the watermelons, both along and across the crystals' long axis. Elbaite is the principal gem mineral that can be and is cut into bicolored and multicolored stones.

Most tourmalines are strongly pleochroic, an important factor in cutting gems; when viewed along the prism axis, the color is deeper or even different from that seen through the side of the crystal. Rare alexandrite-like tourmalines appear yellowish or brownish green in daylight and orange red in incandescent light. Crystals occasionally grow with fluid-filled tubes parallel to the long prismatic direction. If these are sufficiently numerous and narrow, like fibers, chatoyancy will be manifested from a properly fashioned gem.

RIGHT: A superb bicolored elbaite crystal group 10.5 cm (4⅛ in) high from Tourmaline Queen Mine in Pala, California.

OPPOSITE: Faceted elbaites in different colors ranging in weight from 127.70 cts. to 1.27 cts., from various localities.

The crystal structure of tourmaline lacks a center of symmetry; structural elements are much like arrows that point in one direction along the crystal's trigonal axis. As a result, when some crystals are heated, a positive charge develops at one end and a negative change at the other; the reverse occurs upon cooling. This property, the *pyroelectric effect*, was first observed in gem tourmalines.

A comparable electric charging is developed if pressure is applied to the ends of a crystal. This property, *piezoelectricity*, has important industrial and electronic applications. Among common gems, only tourmaline and quartz possess the properties of piezoelectricity and pyroelectricity.

SOME TOURMALINE VARIETIES AND COLORS

Rubellite: Pink to red

Siberite: Violet to red purple

Indicolite: Blue

Paraíba: Electric blue-green (due to copper)

Verdelite: Green

Achroite: Colorless

Note: The new convention is simply to label tourmalines by their color as opposed to their place of occurrence, but old habits die hard, as with the case of Paraíba, only known since the 1980s.

Historic Notes

A long-held belief that tourmaline from Asia was imported by Greece and Rome was recently confirmed when a fine convex intaglio depicting the head of Alexander the Great (now in the Ashmolean Museum in England) was identified as a color-zoned purple-yellow tourmaline. A minute inscription indicates India as the place of origin and the date of its carving as the third or second century BCE. Another much later documented piece is a gold ring of Nordic origin from 1000 CE set with a pink tourmaline cabochon. The crown of Saint Wenceslas in Prague from between 1346 and 1348 displays a large red gem, once thought to be a ruby but actually a tourmaline. Examination of other early surviving jewelry will probably reveal more tourmalines, confirming that they have been used as a gem material for over two thousand years.

The term *carbunculus* was applied to red transparent gems—including ruby, spinel, garnet, and probably red tourmaline—from Pliny's time in the first century CE through

ABOVE: An illustration of a Chinese mandarin from 1800, wearing a badge likely made of red tourmaline.

OPPOSITE: Bicolored elbaite from Brazil—carved rhinoceros, 8.5 cm (3⅔ in) long.

the Middle Ages. Green tourmaline was exported from Brazil to Europe in the early sixteenth century; it was known as Brazilian emerald.

The Chinese valued red and pink tourmaline and made small carved tourmaline ornaments for headdresses and girdles, and certain mandarins wore tourmaline badges or buttons on their caps to signify their rank. Gustavus III of Sweden chose an incredibly large "ruby" as a gift to Catherine the Great when he was on a state visit to Russia in 1777; it is, in fact, a stunning red Burmese tourmaline carved in China.

In 1703, a packet of stones from Sri Lanka labeled "*turmali*" or "*toramalli*" (variations of a Sinhalese word applied to any unidentified yellow, green, or brown stone and meaning "something little out of the earth") arrived at a Dutch lapidary. According to one story, children playing with some of the "pebbles" outside a gem worker's shop noted that when warmed by the sun, the little stones attracted ashes and straws much as a magnet attracts iron filings. So the stones were called *Aschenstrekkers*, or "ash-drawers." (As they are natural dust collectors, the Museum's tourmalines require

81

frequent cleaning as a result of their daily heating by lights in the exhibit cases.) This discovery of tourmaline's pyroelectric property set off a spate of investigations, resulting in observations that only certain gemstones with various colors possessed this property. Finally, in 1801, all the information came together with the recognition of the tourmaline "family."

During the eighteenth century, the principal sources of the mineral were Burma (Myanmar), Russia, Sri Lanka, and Brazil, but late one afternoon in 1820, two Maine schoolboys, Elijah L. Hamlin and Ezekiel Holmes, happened on a brilliant green crystal sparkling in the roots of an overturned tree as they returned from a hike on Mount Mica. The stone was identified as tourmaline, and, starting in 1822 with Mount Mica and later at other Maine locations, mines were opened that could provide sufficient red and green tourmaline to create a market—and value. With the Maine discoveries and several in California, the United States became for a time the world's major supplier of tourmaline. Among the purchasers was Tz'u-hsi (1835–1908), the dowager empress of China, who sent personal emissaries to California to purchase the favored red variety of the gem.

GEMSTONES FREQUENTLY CONFUSED WITH TOURMALINE

These include topaz, beryl (emerald, aquamarine, morganite, and golden beryl), spodumene (kunzite and hiddenite), peridot, andalusite, and apatite. Dark green synthetic spinel is sold as synthetic tourmaline.

Elbaite tourmaline crystals from Mount Mica in Maine. The green tourmaline (upper left) is the original Maine find. Elijah Hamlin had it set in a watch charm bearing the inscription "Primus." It was given to the Museum by his great-granddaughter, K. B. Hamlin. The largest crystal is 5.3 cm (2 in) long.

Legends and Lore

Without many centuries of associated history, the newcomer tourmaline gems do not appear in lore. Even George F. Kunz, popularizer of tourmaline and the man who introduced the stone to Tiffany & Co., opposed acceptance of the gem as the alternate birthstone for the month of October on that account.

With the recent revival of mysticism, tourmaline has become a favorite of New Age adherents, who believe the mineral's pyroelectric and piezoelectric properties produce powerful amplification of psychic energy and neutralization of negative energies.

Occurrences

Tourmaline is a relatively common mineral, the most common boron-bearing silicate. Gem tourmaline is virtually restricted to pegmatites—rich in volatile elements like boron, beryllium, and lithium. Pegmatites yield not only crystals of elbaite but other gem minerals that contain these elements—gemstones such as spodumene and beryl.

Brazil is the world's major source of tourmaline. The pegmatitic region in the eastern part of Minas Gerais yields green, pink, red, and watermelon stones. Tourmaline is also produced in the states of Bahia, Paraíba, and Rio Grande do Norte. The United States also has ranked high as a tourmaline supplier. The 400 pegmatitic dikes of Mesa Grande in San Diego County, California,

produced 120 tons of gem tourmaline between 1902 and 1911. Production reached its peak in 1910; however, with increased Brazilian supply and the fall of the last Chinese dynasty in 1912, California elbaite lost its markets, and many mines closed in 1914. Mining has been reactivated within the last forty years. Most California elbaite is pink and pure in color, but it lacks the exceptional clarity of Maine elbaite. Pegmatites in eastern Maine, such as those at Newry, have

been sporadically productive of the state's gem—tourmaline. Connecticut was a source at the beginning of the twentieth century. Important sources in recent years include Afghanistan, Madagascar, Mozambique, Nigeria (the last two in particular for Paraíba-type color), and Zambia.

Evaluation

Purity and intensity of color and clarity are the most important qualities to consider. The most valued tourmalines are the Paraíba electric blue-green, followed by raspberry red, medium-dark emerald green, and intense blue. Bicolored and multicolored gems are next in value. Cat's eye gems are valuable if the eye is well defined and the fibrous cavities causing it are not coarse. During the past thirty or so years, tourmaline jewelry has been in great demand. Tourmaline crystals are also avidly sought by mineral collectors for the splendor of their colors and forms. George Kunz's faith in his "drop of green light" has been amply justified.

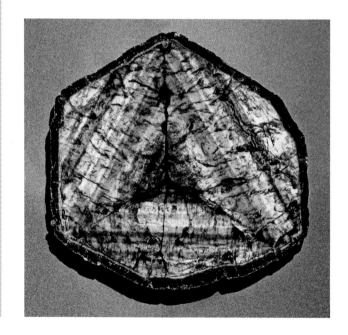

OPPOSITE: Pink elbaites from Pala, California—a crystal 10.0 cm (4 in) long and a 419.50-ct. cut stone.

ABOVE: A cluster of bicolored elbaite crystals 23.0 cm (9 in) long from Nuristan, Afghanistan.

RIGHT: A polished slice of color-zoned liddicoatite, 13.2 cm (5¼ in) wide, from Madagascar. Liddicoatite is an uncommon tourmaline species named in 1977 after gemologist R. T. Liddicoat.

ZIRCON & PERIDOT

Zircon

In the 1920s, a new blue gemstone suddenly appeared on the market. Endowed with spectacular brilliance, it was an immediate hit. The gems were zircons, normally brown to green—but not blue. George F. Kunz, the legendary Tiffany gemologist, immediately suspected trickery; not only were these extraordinary stones available in abundance but available all over the world! Upon Kunz's behest, a colleague made inquiries during a trip to Siam (Thailand) and learned that a large deposit of unattractive brown zircon had stimulated color-improvement experimentation by local entrepreneurs. Heating in an oxygen-free environment had turned the drab material into "new" blue stones, which were sent to outlets worldwide. When the deception was revealed, the market simply accepted the information, and the demand for the new gems continued unabated.

ZIRCON DATA

Zirconium silicate: $ZrSiO_4$

Crystal symmetry: Tetragonal

Cleavage: None, but brittle

Hardness: 7.5

Specific gravity: 4.6–4.7

R.I.: 1.923–2.015 (high)

Dispersion: High

Note: Zircon is not cubic zirconia, a synthetic diamond simulant—zirconium oxide, ZrO_2.

OPPOSITE: Round brilliant-cut zircon from Thailand, weighing 208.65 cts., is the largest known blue zircon on public display.

Properties

Superior brilliance; good dispersion, or fire; clarity; and a breadth of colors stand out as zircon's fine gemstone qualities. Natural zircons range from colorless to pale yellow or green when they initially crystallize. The colors are a result of minor amounts of thorium and uranium that replace zirconium in the crystal structure. But with geologic time, uranium and thorium emissions cause radiation damage, which can be so severe that the original structure is obliterated. A glasslike substance develops with colors ranging from red to brown, orange, and yellow. Heat treatment can restore the structure and color or create new colors, including yellow, blue, and colorless. Colorless zircons imitate diamond's optics better than any other gemstone mineral; their refractive indices approach diamond's, and the dispersive fire is nearly as good.

Many zircons are very brittle; a slight knock will remove a corner or even split the stone. This fragility is the result of internal stress either from radiation damage or from heat treatment. Zircon crystals are distinctive because of their square cross section and pyramid terminations owing to the tetragonal symmetry.

Historic Notes

The Arabic words *zar* and *gun* mean "gold" and "color" and may be the source of the word we use. The terms *hyacinth* and *jacinth* were used in Europe for reddish brown and orange red stones and applied to zircons and other minerals with similar color.

The gem was in use in Greece and Italy as far back as the sixth century CE. Gem zircons were marketed as diamonds sometime after faceting began in the fourteenth century. Colorless zircon was mined at Le-Puy-en-Velay in France in 1590 and sold as Diamond of France. Later, Sri Lankan colorless zircon was sold as Matura diamond (named for the locality where it was found).

Reddish brown zircon became moderately popular in Europe during the nineteenth century, but currently the most commonly used zircons are light blue, golden brown, and colorless.

Legends and Lore

The zircon is among the gems utilized in the sacred, mythological "wish-fulfilling" Kalpataru tree of Hindu religion. The tree was described by nineteenth-century Hindu poets as a glowing mass of precious stones, including sapphire, diamond, and topaz. Green zircon represented the tree's foliage.

As an amulet for travelers, zircon (hyacinth or jacinth) protected its wearer from disease and injury, ensured good sleep, and guaranteed a cordial welcome everywhere, according to the eleventh-century writings of Marbode. Five centuries later, the stone rendered its owner prudent in practical matters (and thus promised financial success) and guaranteed that he or she would never be struck by lightning, according to Girolamo Cardano. By the seventeenth century, belief in the magical properties of gems in general appeared to wane.

Anselmus de Boodt declared that gems cannot of themselves produce supernatural effects. Nevertheless, he believed in the zircon's power to prevent plague.

OPPOSITE: A detail of a carved relief at Borobudur temple on Java, Indonesia, celebrating the sacred, gem-adorned Kalpataru tree, whose leaves were said to be of green zircon.

Occurrences

Zircon is a common minor constituent of igneous rocks, particularly granites, and, to a lesser extent, of metamorphic rocks. Gemstone crystals are rare and found mainly in pegmatites or in fissures. Zircons concentrate in alluvial and beach deposits.

The Chanthaburi area in Thailand, the Palin area in Cambodia, and the southern part of Vietnam near the Cambodian border are the major sources of zircon. It occurs as waterworn pebbles in gem gravels that are seldom more than 10 feet deep. Bangkok is the world's cutting and marketing center for zircon. Blue, colorless, golden yellow, orange, and red stones, almost all of them heat-treated, are exported from there.

Sri Lanka, where zircon is also found in gem gravels, is the next most important source. Other occurrences are in Myanmar, Tanzania, France, Norway, Australia, and Canada.

Evaluation

Color and clarity are the most important considerations in evaluating zircon. The most primary and rare color is red; next is pure, intense blue and sky blue. Colorless, orange, brown, and yellow are less valued. Any visible flaw diminishes the value substantially.

Most zircons have been heat-treated. The color of some of the heat-treated stones may change and the brittleness of some treated gems is also a negative factor. The beauty of this gem—expressed in the variety of its colors and in its clarity, brilliance, and fire—makes it popular today; it is also reasonably priced in comparison with most other gems.

Zircons from Sri Lanka and Thailand, ranging in weight from 7.76 to 40.19 cts.

Peridot

For three millennia, a small, desolate, and forbidding island in the Red Sea has been exploited for the gemstone peridot. Almost nothing grows on this spot of land; there is no fresh water, and the brutal heat relents only in the middle of winter. From the port of Râs Banâs in Egypt, small boats are still used to cross the more than 30 miles of shark-infested water to reach the island. The beaches near the deposits are green with tiny gem crystals. March up Peridotite Hill through the ancient diggings, and you find fissures lined with complete and fractured gem crystals measuring from millimeters to several centimeters. The island is Zabargad, Arabic for "peridot"—the gemstone variety of olivine. For peridot, Zabargad has been the historic and illustrious source.

PERIDOT DATA

A variety of forsterite, Mg_2SiO_4, which with fayalite, Fe_2SiO_4, constitutes a complete series (solid solution) in the olivine group of minerals.

Magnesium iron silicate: $(Mg,Fe)_2SiO_4$

Crystal symmetry: Orthorhombic

Cleavage: Imperfect in two directions

Hardness: 7

Specific gravity: 3.22–3.45

R.I.: 1.635–1.690 (moderate)

ABOVE: Peridot crystal 4.1 cm (1⅝ in) long and a 10.92-ct. cut stone, both from Zabargad Island, Egypt.

Properties

Olive to lime-green color is the most important quality for peridot. This characteristic color is caused by iron; the color saturation increases with iron content; a brownish tinge results from oxidation—minor change of ferrous iron (iron with valence of two) to ferric iron (valence of three). Most peridot is about 90 percent forsterite and the rest fayalite. The transparent gemstone has reasonably good properties: moderate durability and brilliance with a slightly greasy-looking luster.

Historic Notes

The Egyptians fashioned peridot beads as early as ca. 1580 BCE. Second- and first-century BCE writings of the Greek geographers Strabo and Agatharchides described Zabargad and its mining operations. In the third and fourth centuries CE in Greece and Rome, the gemstone was used for intaglios, rings, inlays, and pendants.

During the Middle Ages, the Crusaders brought peridot back to Europe; some of these gems are preserved in European cathedrals. Peridot was highly prized during the latter part of the Ottoman Empire (1300–1918). Turkish sultans amassed the world's largest collection of the gems. In Istanbul's Topkapi Palace Museum, there is a gold throne decorated with 955 peridot cabochons ranging up to an inch across, peridots in turban ornaments and on jeweled boxes, and literally thousands of loose peridots.

When the term *peridot* was first used is uncertain; French jewelers used it long before French mineralogist R. J. Haüy (1743–1822) applied it to the mineral. Yellow-green peridot is sometimes called *chrysolite*, a term deriving from the Greek words meaning "gold" and "stone" and confused with topaz and perhaps chrysoberyl. During the nineteenth century, the peridot became popular in both Europe and the United States, and production on Zabargad was active during the first half of the twentieth century.

Legends and Lore

Ancient Egyptians called peridot the "gem of the sun." An early Greek manuscript on precious stones tells us that peridot bestows royal dignity on its wearer. Another belief was that the stone would protect its owner from evil spirits. In order to do so, the gem must be pierced, strung on the hair of an ass, then tied around the wearer's left arm, a procedure outlined by Marbode. A thirteenth-century English manual states that if a torchbearer, sign of the sun, is engraved on the gem, it will bring wealth to its owners.

A peridot-incrusted lantern hanging in the Topkapi Palace (1459) in Istanbul.

northern Pakistan, peridot was extracted from narrow cavities in peridotite in the early 1990s and became only the third source of large peridot crystals; production has been sporadic owing to political instability and high altitude. Other occurrences are in Minas Gerais in Brazil, Sunnmøre in Norway, China, and Kenya.

Evaluation

The greener the peridot, the higher its value. A tinge of brown diminishes its price, and any flaws make the stone undesirable. Usually, the price per carat does not increase with size.

Occurrences

Forsterite is common in basalts and predominant in peridotite rock, but large unfractured peridot crystals are rare. At Zabargad, the peridotite is cut by coarse veins that are like a pegmatite. Zabargad is presently inactive, awaiting better times in the Middle East.

The major source of peridot for stones of modest size has been peridotite on the San Carlos Indian Reservation in Arizona. The stones rarely exceed five carats. The other historic source of large masses of fine-quality peridot is at Pyaung Gaung at the northern limit of the Mogok Stone Tract in Myanmar. In the remote area of Sapat, Kohistan, in

GEMSTONES FREQUENTLY CONFUSED WITH PERIDOT

These include tourmaline, green zircon, green garnets, chrysoberyl, moldavite (a tektite—natural glass), and sinhalite.

ABOVE: A 164.16-ct. peridot from Myanmar (top); the other gems are 95.19 and 61.55 cts., both from Zabargad.

TURQUOISE &
LAPIS LAZULI

Turquoise

Turquoise is a gemstone with two probable firsts—first to be mined and first to be imitated. Indirect evidence suggests that the Wadi Maghara and Serabit el-Khadem mines on the Sinai Peninsula were in production before 3100 BCE. Egyptian turquoise beads dating to 4000 BCE have been found at al-Badari. Surviving records from the time of King Semerkhet (ca. 2923–2915 BCE, during the First Dynasty) document extensive mining operations that employed thousands of laborers and continued until about 1000 BCE.

By 3100 BCE, either supplies were not meeting demand or a cheaper substitute was desired, because imitations (soapstone glazed blue and green—a form of faience) are found as artifacts of this period.

TURQUOISE DATA

Copper aluminum phosphate:
$CuAl_6(PO_4)_4(OH)_8 \cdot 4H_2O$

Crystal symmetry: Triclinic (normally cryptocrystalline)

Cleavage: Not observed in massive gemstone form

Hardness: 5–6

Specific gravity: 2.6–2.8

R.I.: 1.62 average

OPPOSITE: A 93.98-ct. high-crowned cabochon from Iran, and a 90.20-ct. cabochon of turquoise with spiderweb matrix from Santa Rita, New Mexico.

Properties

Color is turquoise's superlative gem property; the mineral's other properties are less than ideal. The gemstone usually forms in aggregates of submicroscopic crystals that make it opaque. Turquoise is relatively soft and subject to scratching. Its porosity makes it discolor by absorption of oils and pigments, and friability can lead to easy breakage; only the most compact varieties resist these tendencies. The sky-blue color is intrinsic, a result of copper. Iron in turquoise leads to greener tones. Ochre or brown-black veining is common, the result of oxide staining or inclusion of adjacent rock fragments during turquoise's formation.

Historic Notes

The name *turquoise* did not come into use until the thirteenth century. Pliny used the term *callais*, derived from the Greek *kalos lithos*, meaning "beautiful stone." Purchased by Venetian merchants in Turkish bazaars for European trade, the blue stone was called "pierre turquoise" by the French recipients, meaning "Turkish stone."

The first uses of turquoise were in Mesopotamia (Iraq), where beads dating from about 5000 BCE have been found. Turquoise is Iran's national gemstone. It has decorated thrones, daggers, sword hilts, horse trappings, bowls, cups, and ornamental objects. High officials once wore turquoise seals decorated with pearls and rubies. Large stones were embellished with gold scrollwork to hide the imperfections. After the seventh century CE, turquoise pieces decorated with passages from the Koran or Persian proverbs in incised gilt characters were treasured as amulets. Turquoise has been the most cherished gem in Tibet as well, with a role comparable to that of jade in China.

Turquoise was frequently set in Siberian jewelry of the sixth and fifth centuries BCE and in pieces from southern Russia of a slightly later date. The ancient Greeks and Romans engraved turquoise for ring stones and pendants and also used it as beads, but certainly it was not one of their favorite stones. In Europe, it became more popular during the Middle Ages for decoration of vessels and the covers of manuscripts. Popularity of the stone for personal adornment grew during the Renaissance; by the seventeenth century, "no man considered his hand well-adorned" unless he wore a turquoise, according to Anselmus de Boodt. In the following centuries, turquoise remained popular, gracing royal crowns in addition to modest jewels as accent stones. In Europe, it has been the most popular opaque gemstone.

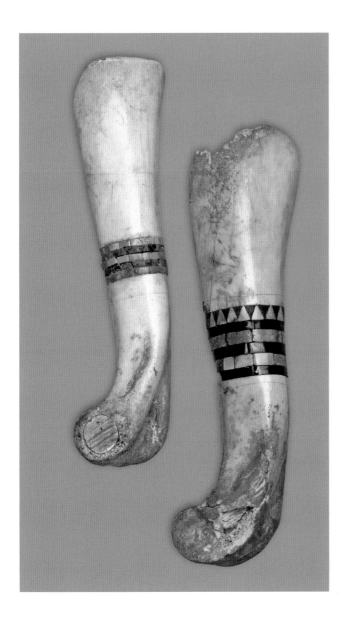

ABOVE: Bone scrapers inlaid with turquoise and jet, 15 cm (6 in) long, found during the 1896 Hyde Expedition in Pueblo Bonito, Chaco Canyon, New Mexico.

OPPOSITE: The turquoise Buddhist lion was carved from Tibetan material and is 6.1 cm (2⅜ in) long.

The gem's history in the Americas began about 1,000 years ago with the initiation of turquoise mining at Mount Chalchihuitl in Cerrillos, New Mexico. Native Americans, using hand tools, quarried the entire mountain; all that remains on the north side is a pit about 200 feet across and up to 130 feet deep.

Turquoise has been found in burial sites in Argentina, Bolivia, Chile, Peru, Mexico, Central America, and the southwestern United States. The Incas carved beads and figurines and crafted beautiful turquoise inlays. The Aztecs used turquoise in mosaic pendants and ritual masks. The Zunis, Hopis, Pueblos, and Navajos all made magnificent necklaces, ear pendants, and rings. At Pueblo Bonito in northwestern New Mexico, nearly 9,000 beads and pendants were found near a single skeleton. All told, 24,932 beads were found in these burial sites.

Legends and Lore

In Persia (Iran), one who could see the reflection of a new moon on a turquoise was certain to have good luck and be protected from evil. Hindus had a comparable belief: if an individual could look at a new moon and immediately after at a piece of turquoise, great wealth would surely follow. To the Navajos, a piece of turquoise thrown into a river (while a prayer to the rain god was being spoken) would ensure the blessing of rain. A turquoise attached to a gun or bow would guarantee accurate aim, according to Apache lore.

The belief that turquoise would protect its owner from falling, especially from a horse, was first recorded in the thirteenth century. This virtue is traceable to the use of turquoise in Persia and Samarkand as a horse amulet. Legend also has it that, by changing color, turquoise reveals a wife's infidelity.

Occurrences

Turquoise crystallizes as veins and nodules near the water table in semiarid to arid environments. Its chemical stockpiles are the adjacent rocks, which are leached by rain and groundwater; thus turquoise is often associated with weathered igneous rocks containing primary copper minerals.

Before World War I, turquoise production from nearly one hundred mines was Iran's most important industry. Following World War II, output declined but was revived after the revolution. Turquoise is found in Nevada, Arizona, Colorado, New Mexico, and California, the primary producers of turquoise today. Much of the turquoise is a by-product of copper mining. Most American turquoise is light in color, porous, and chalky, usually with matrix, and only 10 percent of the turquoise mined is of gem quality. Other occurrences are in Armenia, Kazakhstan, China, Australia, Israel, Tibet, Mexico, and Afghanistan.

Evaluation

The intensity and evenness of color and quality of polish affect the value of turquoise. The very rare, intense sky blue (robin's egg blue) is most desired. Turquoise with matrix is generally less valuable than stones without it. Of matrix turquoise, the spider web variety is the most valuable.

A fine polish is possible only with stones that are hard, relatively nonporous, and compact. Very pale and chalky turquoise is sometimes impregnated with oil, paraffin, resins, glycerin, or sodium silicate to enhance its color and ability to take a good polish. Occasionally, this turquoise is sold as stabilized turquoise or turquolite. Some turquoise is even painted on the surface with blue dye and then coated with clear plastic. A product consisting of powdered turquoise in epoxy is also widespread. Much of the turquoise on the market has been treated in these ways, and some may change color. The admonition "buyer beware" certainly befits purchasers of turquoise.

GEMSTONES CONFUSED WITH TURQUOISE, AND IMITATIONS AND SYNTHETICS

Chrysocolla, chrysocolla quartz, odontolite (a naturally stained fossilized bone), variscite, and malachite are easily confused with turquoise.

Turquoise may be imitated with glass, porcelain, plastic, enamel, stained chalcedony, dyed howlite, blue-dyed and plastic-treated marble, and doublets.

Artificial products are sold with names such as Viennese turquoise, Hamburger turquoise, and Neolith. Synthetic turquoise has been produced and marketed in France since 1970.

A polished variscite sphere 7.5 cm (2¹⁵⁄₁₆ in) in diameter, from Fairfield, Utah. Variscite is often mistaken for turquoise.

Lapis Lazuli

Lapis lazuli is probably the original blue gemstone and has an ancient history of exploitation in Afghanistan. The gemstone has been so important to Afghanistan that lapis figured in U.S. foreign policy. In 1985, during Senate Armed Services Committee hearings on the Soviet war in Afghanistan, some testimony indicated that lapis lazuli was an important source of cash for the mujahideen to buy weapons to battle the Soviet-backed regime. In fact, the Kabul government was also trying to raise cash by selling quantities of the blue stone. The result for the market had been an unusual abundance of lapis available during the Russian war. In the years of the war against Al-Qaeda and the Taliban, supply has varied, but Afghan lapis is still generally available. Nonetheless, fine and uniform blue color with only occasional golden flecks that distinguish the most desirable lapis lazuli is not very abundant.

LAPIS LAZULI DATA

A rock composed principally of the mineral lazurite [a sodium alumino-silicate containing sulfur, chlorine, and hydroxyl—$(Na,Ca)_8(Al_6Si_6)O_{24}[S_2,(SO_4)] \cdot nH_2O$ (n<1) with variable amounts of pyrite (the brassy flecks) and white calcite

Cleavage: Not relevant for a rock

Hardness: 5–5.5

Specific gravity: 2.7–2.9

R.I.: About 1.5 (opaque)

ABOVE: Polished lapis lazuli boulder from Afghanistan, 13.5 cm (5¼ in) high.

Properties

As lapis lazuli is opaque, the most important qualities are color and a moderate durability. Lazurite is not very hard, but fine-grained lapis is reasonably tough. The intrinsic blue color of lazurite is caused by sulfur, an interesting and unusual case of a nonmetallic element yielding a strong color. Lapis was ground into the pigment ultramarine until a synthetic equivalent was developed in 1828.

Historic Notes

In Egypt, carved lapis beads, scarabs, pendants, and lapis-inlaid jewelry date prior to 3100 BCE. The stone was esteemed as a gem and amulet. Ground into powder, it was used as a medicine and a cosmetic—the first eye shadow.

The tomb of Queen Pu-abi (2500 BCE), in the city of Ur in Sumer, contained adornments rich with lapis—three gold headdresses, two bead necklaces, a gold choker, a silver pin, and a gold inlaid ring.

During the time of Confucius (ca. 551–479 BCE), the Chinese carved lapis hair and belt ornaments. As

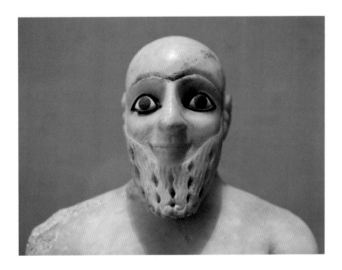

early as the fourth century BCE, the Greeks used lapis for carving scarabs and scaraboids, and it was described by Theophrastus. In Rome, lapis was fashioned into intaglios, plain ring stones, beads, and inlays. The ancient Greeks and Romans used the term *sapphirus*; *lapis lazuli* did not come into use until the Middle Ages. Lazulus means "blue stone" in Latin and derives from the ancient Persian *lazhuward*, meaning "blue" and the Arabic *lazaward*, meaning "heaven," "sky," or simply "blue in general."

The stone was a favorite material for carving objets d'art during the Renaissance in Europe. When Catherine the Great was told that lapis had been discovered near Lake Baikal, she ordered that mining be commenced immediately. In the following year, 1787, the empress decorated a room in her palace in Tsarskoye Selo (now Pushkin) with the stone. Sections of walls, doors, fireplaces, and even mirror frames were made of lapis.

Today, lapis is favored for beads, ring stones, and pendants and is a preferred stone for men's jewelry.

Legends and Lore

To the Buddhists, lapis brought its owner peace of mind and equanimity and dispelled evil thoughts. In *De Materia Medica* (ca. 55 CE), Greek physician and pharmacologist Dioscorides recommended lapis as an antidote for the bite of a poisonous snake. By the thirteenth century, broadened curative powers were attributed to lapis; Albertus Magnus advised using lapis for intermittent fever and melancholy in his mineralogical treatise.

LEFT: Lapis lazuli eyes grace the Louvre museum's statue of Ebih-II (ca. 2400 BCE), the superintendent of the ancient Syrian city-state of Mari.

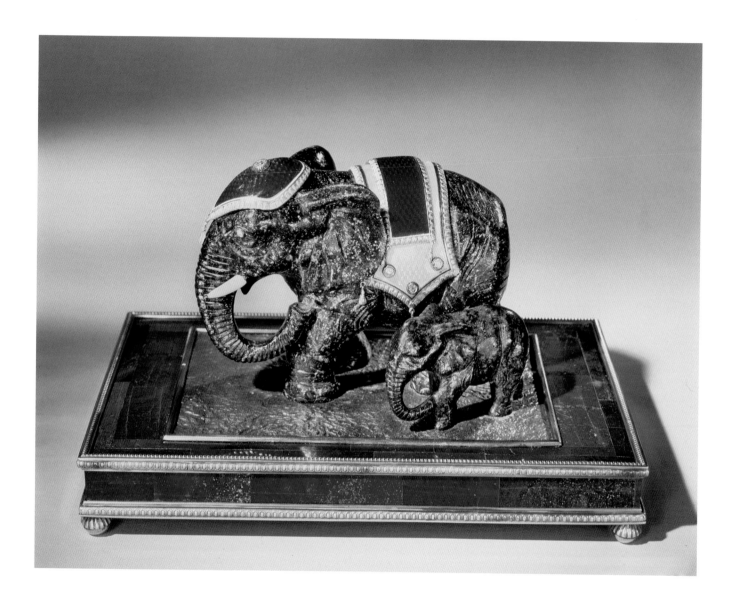

ABOVE: Russian lapis lazuli carving decorated with sterling silver, yellow gold, red and yellow enamel, and small rose-cut diamonds. It measures 16 cm (6¼ in) across, and the base is stamped by the famous Fabergé workshop.

Occurrences

Lapis is a rare metamorphic rock produced by the interaction of granitelike magma with marble. In addition to Afghanistan, Chile is a major source of the gemstone. The most productive mine is in the Andes in Coquimbo Province, north of Santiago. It was worked by the Incas in

101

pre-Columbian times and continues in production today. A less important source is near Antofagasta. Chilean lapis usually contains large amounts of calcite, although mining of better-grade material has been reported recently. Russia produces the stone from mines near Lake Baikal, as does Tajikistan near Khorog in the Pamir Mountains. Small sources exist in the Mogok Stone Tract, Myanmar; Pakistan; Mongolia; Canada and Italy. In the United States, lapis is produced in Colorado and California.

Evaluation

The quality, purity, and evenness of color largely determine the value of lapis. The most desired color is intense violet blue. Stones without inclusions of pyrite or calcite are also most desirable. Lapis with inclusions of pyrite is more valuable than that with inclusions of calcite. Often the white calcite inclusions are disguised with paraffin treatment, which may include the use of blue dye. (Dye can be detected by rubbing the stone with cotton dipped in acetone or finger-nail-polish remover; the blue color comes off on the cotton. In fact, this test works for most dyed gems.) The minerals in lapis have unequal hardness and polish differently. Only superior, inclusion-poor lapis can be polished to a smooth, even luster.

GEMSTONES CONFUSED WITH LAPIS LAZULI, AND SUBSTITUTES AND IMITATIONS

Sodalite, azurite, lazulite, and dumortierite may be confused with lapis lazuli.

The most common substitute for lapis is blue-dyed chalcedony, sold as German lapis and Swiss lapis. Lapis has been imitated with synthetic spinel with gold forced into surface holes to simulate inclusion of pyrite (fool's gold).

OPPOSITE: Lapis lazuli Chinese junk, 15.2 cm (6 in) high.

OPAL

The sheer beauty of opal outweighs its disadvantageous physical properties. Rainbow colors pour out of well-lit precious opals, but the gems are easily scratched and so are a poor choice for exposed ring settings. Opals are mechanically fragile and notoriously difficult to set in jewelry; a slight blow or a rapid change in temperature may shatter an opal. The problem is that the gem contains water, which—depending on the opal and its source—may evaporate and leave the opal slightly smaller, stressed, and covered with cracks. Opals need our protection; worn close to the body, they are safe from abrasion, are kept at an even temperature, and receive some body moisture so that they do not lose water.

OPAL DATA

Hydrated silica: $SiO_2 \cdot nH_2O$

Crystal symmetry: Largely amorphous

Cleavage: None, but brittle

Hardness: 5.5–6.5

Specific gravity: 1.98–2.25

R.I.: 1.43–1.47 (low)

Special optical property: Diffraction—play of colors

OPPOSITE: The 215.85-ct. Harlequin Prince, found in Australia.

precious opal contains many of these organized zones that display diffraction colors, whereas common opal may be colored but does not show a play of colors. Precious opal is usually cut as a cabochon or carved, but some fire opal is sufficiently transparent to be faceted.

Opals without play of colors, so-called common opals, are also used as gems. Fire opal may or may not have the play of colors, but rather an intense red-orange body color with near transparency that is stunning. Yellow and brown translucent opals are available at modest cost. Picture opals have veins or inclusion trains resembling abstract paintings. *Potch* is a term for common opal, used by miners seeking precious opal.

Properties

The characteristic feature of precious opal is its play of colors; pure colors can be seen in rapid succession when the gem is moved about. By some standards, opal is not a mineral because it does not have an extended crystal structure. Opal is made up of submicroscopic silica spheres bonded together with more silica and water. The lower the initial amount of water in the opal, the better its physical properties. Loss of water or change in temperature causes strain that is relieved by cracking, known as crazing. Also, opal is soft, and its density and refractive index are low.

If the minute spheres in opal are uniform in size and packed into a regular array, they can scatter light in various colors (by diffraction) determined by the size of the spheres and the opal's orientation. Gem or

TOP LEFT: Opal carving of a leaf from rough found in Stuart Range, South Australia, measuring 6.3 cm (2½ in) in length.

ABOVE: Opals from Mexico ranging in weight from 4.72 to 31.70 cts.

PRECIOUS OPAL
BODY COLORS AND THEIR CAUSES

Precious opals are generally defined by manifesting a play of colors:

Black: A black or dark background (or body) in gray, blue, or green—dark inclusions

White: A white background—internal boundaries or fluid inclusions

Water: A transparent or colorless stone—few or no inclusions

Fire: A transparent stone (play of color may or may not appear) or translucent yellow, orange, red, or brown— ochre-colored iron oxide in inclusions

Boulder opal: Opal, sometimes precious, filling cracks and cavities in a rocky or ironstone matrix

Opal classification by size and pattern of color patches:

Pinpoint or pinfire: Small patches, close together

Harlequin: Larger, more angular patches that resemble the diamonds of a harlequin pattern

Flame: Red streaks and bands that cross the surface like flames

Flash: Flashes of color that appear and disappear when the stone is moved

Historic Notes

Opal derives from the Sanskrit *upala* and the Latin *opalus*, meaning "precious stone." In his *Natural History*, Pliny described fine opals: "In the opal you will see the refulgent fire of the carbuncle, the glorious purple of amethyst, and the sea green of the emerald, and all these colors glittering together in incredible union."

The oldest opal mines were at Czerwenitza (Červenica), now in Slovakia (formerly Hungary), northeast of Košice. Archival evidence indicates that the mines were worked in the fourteenth century, but there are indications that they were in operation much earlier—perhaps the source of opal for Rome. Production of semitranslucent milky-white stones with play of color continued until 1932.

Mexican fire opal was known to the Aztecs and was introduced in Europe by the Spanish conquistadors early in the sixteenth century.

Shakespeare referred to opal as "the queen of gems" in *Twelfth Night*. Opal's prominence declined during the nineteenth century, when the stone was associated with bad luck. Many believe that Sir Walter Scott was responsible. In his 1829 novel, *Anne of Geierstein*, the heroine's demonic ancestor, Hermione, died soon after a drop of holy water touched the enchanted opal that adorned her head and quenched its fire.

Scene from Sir Walter Scott's *Anne of Geierstein* depicting the opal-wearing Hermione, early nineteenth century; possibly a sketch for an oil painting by British painter William Long.

Black opal was first found in Australia in 1887. Queen Victoria helped popularize it and white opal by giving opal jewelry to all of her children. Opal was a favorite stone of René Lalique (1860–1945), the most talented artist-jeweler of the Art Nouveau movement. He designed opal jewelry for Sarah Bernhardt (1844–1923) among others.

Black and white opals are currently among the most popular gems.

Legends and Lore

The Romans considered opal a stone of love and hope; however, Pliny's only description of an opal, one the size of a walnut, suggests it was another stone such as an iris quartz, as opals of that size were not available. The Arabs believed it fell from heaven in flashes of lightning. Marbode wrote of an opal talisman of the House of Normandy: "It renders the wearer invisible, enabling him to steal by day without risk of exposure to baneful dews of night." Thus the opal became known as the talisman of thieves and spies. Anselmus de Boodt summarized: "Opal possesses the virtues of all gems as it displays their many colors." According to an Australian legend, the stars are governed by a huge opal that also controls gold in the mines and guides human love.

OPPOSITE: Pendant 4.5 cm (1¾ in) in length, set with Australian opals, chrysoberyl, sapphires, topaz, demantoid garnets, and pearls; Tiffany & Co. (designed by L. C. Tiffany), ca. 1915–25.

RIGHT: Opal Indian head carving measuring 4.2 cm (1⅘ in) in length. The rough opal is from Mayneside, West Queensland, Australia.

Occurrences

Opal is formed in cavities and cracks in near-surface volcanic rocks or as replacements—thus making fossils—of shells, bones, and wood in or near sedimentary volcanic ash by percolating water that dissolves silica and then precipitates opal.

Australia still produces a majority of the world's play-of-color opal. Lightning Ridge, New South Wales, is the primary source of black opal. Coober Pedy in South Australia, where opal was discovered in 1915, is the opal capital of the world. The aboriginal name for the location, Kupa Pita, means "white man in a hole." The miners and their families all used to live underground, attempting to escape the intolerable climate; some still do. White Cliffs and Andamooka are also important sources. Yowah, in Queensland, is known for its boulder opal.

109

The only commercial source of fire opal is Mexico, principally near Querétaro, where mining began in 1870. Mexican fire opal is softer and lighter than opal from other sources because it contains more water. The states of Jalisco, Hidalgo, Guanajuato, and Nayarit also produce opal.

Opals from the Wollo Province of Ethiopia appeared in the marketplace in 2008 and have grown in abundance since. The deposit, formed in volcanic ash, is large but difficult to mine because of extreme vertical exposures, and the opals vary over a very broad spectrum of types, including spectacular plays of color. Stability of these opals has been raised as an issue for their long-term value.

Opal was discovered at Virgin Valley in Nevada in 1908. This material is very beautiful, but once it is exposed to air, it loses some of its water and cracks. Because no effective method has been discovered to prevent cracking, production remains very limited. Brown common opal and opalized wood are sourced in Oregon and Washington. Other commercial opal sources are in Brazil, Honduras, Indonesia, Madagascar, Peru, and Turkey.

Evaluation

The play of colors is most important to the value of an opal. Fine opals exhibit bright, intense colors. No dead patches should be present. Black opals are more valued than white. Of the types of color patterns, the harlequin is the most valuable. Fire opals without play of color are judged by the beauty of their color and transparency. Red opals are more valued than yellow and brown.

Some white opals from Australia are treated to make them black. After being soaked in sugar solution, they are immersed in sulfuric acid, which carbonizes the sugar, blackening the stone. Some Mexican opals have been turned black by smoke treatment in a mixture of charcoal and cow manure. Plastic and silica-based polymers, with or without dye, have been used to impregnate Brazilian, Mexican, and Idaho opals, significantly improving their appearance.

Opal too thin to be used in jewelry is made into doublets. A thin piece of gem-quality opal is cemented to a piece of common opal, chalcedony, or glass. If an opal doublet is covered with a protective piece of colorless quartz, it becomes a triplet, which is more durable than a doublet.

OPAL IMITATIONS, SUBSTITUTES, AND SYNTHETICS

Glass and plastic are used to imitate opal, but do so only poorly. No natural mineral resembles opal, with the possible exception of labradorite feldspar.

Synthetic opals were first produced commercially in France by P. Gilson in 1972 and are successful.

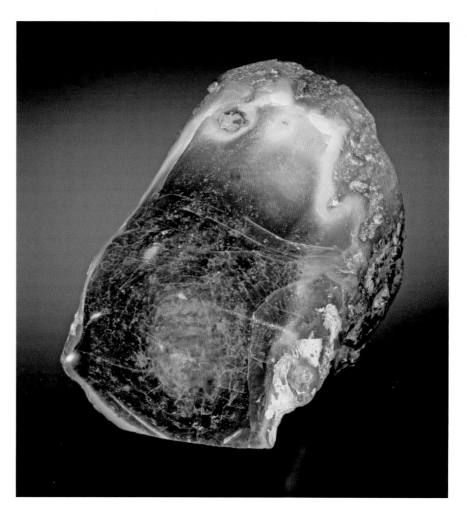

OPPOSITE: An opalized clam weighing 69.00 cts. from Coober Pedy, Australia.

RIGHT: Opalized wood, 9 cm (3½ in) long, from Virgin Valley, Nevada.

FELDSPAR

Minerals of the feldspar group constitute more than half of the Earth's crust, but nature only rarely yields them as gemstones. Remarkable iridescence is the hallmark of the best-known varieties: moonstone and labradorite. The blue shimmer of a fine moonstone forms a subtle but beckoning adornment in soft illumination, whereas labradorite's color flash, like the peacock's tail feathers, is arresting. Of the latter, Ralph Waldo Emerson said in his 1884 essay "Experience": "A man is like a bit of labrador spar, which has no luster as you turn it in your hand, until you come to a particular angle; then it shows deep and beautiful colors."

FELDSPAR DATA

A mineral group forming two distinct compositional series (solid solutions) of alkali aluminosilicates.

Plagioclase Series: $CaAl_2Si_2O_8$ to $NaAlSi_3O_8$

Alkali Feldspar Series: $KAlSi_3O_8$ to $NaAlSi_3O_8$

Crystal symmetry: Monoclinic or triclinic

Cleavage: Two perfect at right angles; imperfect in a third direction

Hardness: 6–6.5

Specific gravity: 2.55–2.76

R.I: 1.518–1.588 (low)

Special optical properties: Light scattering and iridescence

OPPOSITE: A polished labradorite disk measuring 3.2 cm (1¼ in) in diameter, and a 7.6 cm (3 in) long slice, both from Labrador, Canada.

Properties

Iridescence and color are the source of the feldspars' gemstone appeal. They have little brilliance, and crystals are known for their cleavages. The term *feldspar* stems from these cleavages; it means "field spar." The Anglo-Saxon *spar* refers to easily cleaved minerals such as calcite, fluorite, and feldspar.

The iridescence is caused by scattering of light from thin layers in the gemstone; these layers are a second feldspar that develops by internal chemical separation during geologic cooling of an initially single feldspar. Light scattering from the layers results in pure iridescent colors (in labradorite, called "labradorescence" or "schiller") from red to blue, or in a broad spectrum of blue white to yellow white (in moonstone). Peristerite, intermediate in iridescence, gets its name from the Greek *peristera*, meaning "pigeon stone," in allusion to the color of the bird's neck feathers. Sunstone, also known as aventurine feldspar, reflects internal gold spangles.

Visible intergrowths produce translucency or opacity rather than iridescence in most feldspars, and most are usually not strongly colored. Moonstones are colorless to gray or yellow and semitransparent to translucent. Gem orthoclase is transparent and yellow. Labradorite is usually gray and opaque, but rare transparent crystals are occasionally found. Amazonite is an opaque feldspar with vivid green to blue green color. Translucent feldspars are carved or cut *en cabochon*; the rare transparent material is faceted for collectors.

Historic Notes

Amazonite (named after the Amazon River) was widely used in Egyptian, Sudanese, Mesopotamian, and Indian jewelry; some examples date back to the third millennium BCE. The twenty-seventh chapter of the Egyptian Book of the Dead was engraved on this feldspar. A carved scarab and an amazonite-inlaid ring were among the jewels of Tutankhamen (reigned 1333–1323 BCE). Amazonite was treasured by the Hebrews, and it is generally accepted that the third stone in Moses's breastplate was amazonite. In Central and South America, adornments in pre-Columbian times contained amazonite.

Moonstone appeared in Roman jewelry in about 100 CE and even earlier in Asian adornment. The gemstone

LEFT: Labradorites ranging in weight from 2.09 to 3.01 cts. from Clear Lake, Oregon.

OPPOSITE: Amazonite crystal measuring 6 cm (2⅜ in) across from the Lake George area in Colorado and cabochons carved with floral designs, with weights ranging from 18.0 to 29.9 cts., from Amelia Court House, Virginia.

was a favorite of Art Nouveau jewelers, and Cartier and Tiffany creations frequently contained the gem.

Labradorite was used in decoration by the Red Paint People of Maine long before the year 1000. It was "found" by Moravian missionaries in 1770 in Labrador, Canada, and named for the locality.

In the late eighteenth and nineteenth centuries, two important deposits of sunstone feldspar were found in Russia; as a result, the gem had extensive use in Russian jewelry. When sunstone deposits were found in Norway in 1850, the gem became increasingly popular in Europe.

FELDSPAR GEMSTONES

For varieties, the appropriate mineral name is given
in parentheses.

PLAGIOCLASES

Labradorite: Middle of series—colorful iridescence, also
transparent stones in yellow, orange, red, green

Sunstone (oligoclase): Near Na end—gold spangles from
oriented inclusions of hematite

Peristerite (albite): Near Na end—blue white iridescence
(can be mistaken for moonstone)

ALKALI FELDSPARS

Orthoclase: At K end—transparent gemstone, pale to
bright yellow from iron substitution for aluminum

Amazonite (microcline): At K end—yellow green to
greenish blue, opaque; color from natural irradiation of
microcline containing lead and water impurities

Moonstone (sanidine): Near K end—blue white to white
iridescence)

Legends and Lore

Amazonite was a popular amulet among ancient Egyptians.
According to Pliny, the Assyrians considered it the gem of
Belus, their most revered god, and used it in religious rituals.

In India, moonstone was sacred and also had a special
significance for lovers; if they placed it in their mouths when
the moon was full, they could foresee their future. In Europe,
Marbode's eleventh-century lapidary reported that the gem
could bring about lovers' reconciliation. Girolamo Cardano
wrote in the sixteenth century that moonstone could drive
sleepiness away.

Occurrences

Feldspars constitute substantial portions of many igneous
and metamorphic rocks. Gem varieties result from the rare
geologic conditions that produce clean large grains, particularly in pegmatites and ancient deep crustal rocks.

Important localities for amazonite are in India,
Brazil, Quebec in Canada, Madagascar, Russia, South Africa,
and Colorado and Virginia. The best-quality moonstones
came from a dike at Meetiyagoda in southern Sri Lanka;
this source is now exhausted. Moonstones are found in
the gravels of Sri Lanka and in the Mogok Stone Tract of
Myanmar, curiously, as crystals and pods in marble that also
yields ruby and spinel. Also, Chennai (Madras), India, produces a less valuable quality—almost opaque and yellowish,
reddish brown, or grayish blue. Other sources are Brazil,
Australia, and Madagascar. The finest-quality peristerites
are found in Ontario and Quebec and in Kenya. Norway and
Russia continue to be the sources of sunstone, and Labrador
and Finland are the principal source of iridescent labradorite. Transparent, facetable labradorite is found in Mexico
and in Utah, Oregon, California, and Nevada.

Evaluation

Moonstone is the most valuable feldspar. Fine-quality moonstone is semitransparent and flawless and exhibits a broad
blue sheen. Bright-colored amazonite is the most desirable.
To be of fine quality, labradorite must display intense iridescent colors. Spectrolite is a trade name for the Finnish
material. Sunstones that are semitransparent and show a
pleasing reddish or yellow orange glow are the most desirable. A popular imitation, marketed as goldstone, is glass
with copper inclusions.

GEMSTONES CONFUSED WITH FELDSPAR

Moonstone: Quartz, chalcedony, and opal

Amazonite: Jade and turquoise

Labradorite: Opal

ABOVE: Four moonstone intaglios of siblings, by noted engraver Ottavio Negri, ca. 1910, average 2 cm (3¾ in) high. The moonstone is from Sri Lanka. The top center intaglio is enlarged to show details.

JADE

The term *jade* for the ornamental stone most identified with China is a total misnomer. In the sixteenth century, Spanish conquistadors learned of a stone worn by Mesoamericans as an amulet to cure colic and similar maladies. The Spanish called it *piedra de la yjada* (in Latin, *lapis nephrictus*), meaning "stone of the loin," and brought fine examples back to Europe. In translation from Spanish to French, the phrase was misprinted as *pierre le jade*. In the mid-seventeenth century, the New World sources had disappeared, and Europeans forgot the material but not the name; they applied it to the stone of numerous carvings arriving from China. In 1780, geologist A. G. Werner described the traditional Chinese carving material and labeled it *nephrite*, after the Latin term. In 1863, French chemist Augustine Damour chemically analyzed a Chinese carving of a somewhat harder stone and found that it was different from nephrite. He labeled this material *jadeite*, derived from the "original" term *jade*. It was also learned that the nephrite stones were sourced in China proper, whereas the jadeite was sourced from northern Burma (Myanmar). In 1881, Damour discovered that Burmese jadeite and the original Mesoamerican material were mineralogically identical. Nevertheless, the common term *jade* persists for both jadeite and nephrite. To make matters even more complicated, other stones that appear similar or have been used in a similar manner in ancient cultures are also simply called "jade." Such is the confusion with the most important ornamental gemstone.

JADE DATA

Both nephrite and jadeite jade are rocks composed of essentially a single mineral: tremolite in nephrite and jadeite in the other. For rock, crystal symmetry and cleavage are meaningless.

TREMOLITE (NEPHRITE)

Composition: Calcium magnesium silicate
Formula: $Ca_2(Mg,Fe)_5Si_8O_{22}(OH)_2$
Hardness: 6
Specific gravity: 2.9–3.1
R.I.: 1.62 (average)

JADEITE (JADEITE JADE)

Composition: Sodium aluminum silicate
Formula: $NaAlSi_2O_6$
Hardness: 6.5–7
Specific gravity: 3.1–3.5
R.I.: 1.66 (average)

OPPOSITE: Jadeite incense burner, 28 cm (7 in) high, from the reign of Qianlong, China, part of a three-piece altar set; Myanmar jade.

Properties

The special quality that nephrite and jadeite jade share is exceptional durability; nephrite is one of the toughest known substances. Both rarely yield to a hammer blow—a convenient field identification technique (obviously not suggested for art or artifacts). This property means that jade can be carved into remarkably fine and intricate forms with minimal risk of breaking.

Nephrite owes its exceptional toughness to a solid feltlike structure of intergrown microscopic fibrous crystals. Jadeite forms larger prismatic or interfingering crystals that interlock and create a strong network. Both materials accept a fine polish because of their compactness, though nephrite's polished surface often has many small depressions, like an orange peel.

Colors and patterns are quite variable for both nephrite and jadeite. Veins, clots, zoning, and deformation can produce color variations and juxtapositions. Individual colors are due to substitutions of elements in the major constituent mineral or to "contaminant" minerals in the rock. Boulders of both nephrite and jadeite, particularly the green varieties, frequently have a tan- to ocher-colored rind due to oxidation of their constituent iron or impregnation with rust stain during weathering.

ABOVE: Jadeite cabochons from Myanmar (nine smaller stones) and Guatemala (two larger stones), with weights ranging from 6.38 to 28.34 cts. displayed on a 20-kg (45-lb.) jadeite boulder from Myanmar.

JADE COLORS AND THEIR CAUSES

NEPHRITE

White: Essentially pure tremolite, very little iron; sometimes called "mutton-fat jade."

Deep green: Iron "spinach green jade" from Siberia has blotches caused by graphite inclusions.

Creamy brown: The color of this stone—sometimes called "tomb jade"—was once attributed to the action of heat on lime impurities, but research indicates it is the result of reactions between fluids in mummies and jade, both sealed in sarcophagi.

JADEITE JADE

White: Pure jadeite.

Leaf and blue green: Iron.

Emerald green: Chromium; Imperial jade or *fei tsui* is the finest translucent variety.

Lavender: Manganese without the presence of iron.

Dark blue green: Iron in omphacite (a calcium-rich jadeitic pyroxene—approximately $CaNa(Mg,Fe)AlSi_4O_{12}$) and aegirine; once called chloromelanite, a term now invalid in mineralogy. The term omphacite jade is now valid in gemology.

Deep emerald green: Due to substantial amounts of the mineral kosmochlor, $NaCrSi_2O_6$; this jade variety is called *mawsitsit* or *tawmawite*.

Historic Notes

Both jades share a utilitarian beginning. The quality that drew primitive people's attention to them was their singular toughness—in this sense, we can consider them the same. Jade was the raw material for celts (a type of prehistoric ax head), blades, and clubs. One could fashion a thin, strong

edge that retained its sharpness. As cultures developed, jade became a substance valued for its beauty. Thus the concept of jade, regardless of the material or name, became comparable around the world.

Figure of Guanyin, a Chinese Buddhist deity, carved from Burmese jadeite jade in China during the nineteenth century. It is 28 cm (11 in) high without the stand.

Jade has had its longest and most continuous history in China, nephrite being the principal jade used. According to folklore, in the twenty-seventh century BCE the Yellow Emperor, Huangdi, had jade weapons and bestowed jade tablets upon officials to confer rank and authority. In fact, the Chinese symbol for *jade* appears to have been derived from the symbol for *ruler*. Artifacts are known from what became the Liangzhu (3300–2250 BCE) and Hongshan cultures (4700–2900 BCE). By the end of the Zhou Dynasty (255 BCE), carving design had reached maturity, and during the reign of Qianlong (1736–95 CE), technical skill in carving achieved the highest level.

Jade has played a part in almost every aspect of Chinese life—as tools; currency; and awards for statesmen, visiting ambassadors, and war heroes. Some of the earliest records of events are inscribed on nephrite tablets. Ceremonial libation vessels, incense burners, and marriage bowls were carved from nephrite; and it was the material for innumerable personal ornaments, including beads and pendants inscribed with poetry and worn as talismans.

For the Chinese, jade was not only pleasing to the eye but also to the ear and sense of touch. Musical instruments carved from jade have been used for rituals since ancient times, and rounded and well-polished "buttons" were carried in the sleeves for fingering until the end of the nineteenth century.

The earliest use of jadeite jade is documented from the Jōmon culture (ca. 10,500–300 BCE) in Japan. The second group of great jade cultures were the Olmec, Maya, Toltec, Mixtec, Zapotec, and Aztec of Central America, who employed jadeite jade. Carbon 14 dating of wood fragments found with jadeite artifacts in Mexico provides a date of use by the Olmecs around 1500 BCE. Jadeite was typically carved in the form of jaguars and decorated cells for ceremonial purposes. The numerous jade artifacts found in tombs include earplugs, diadems, necklaces, pendants, bracelets, masks, and statues of the sun god. The Spanish conquistadors

completely destroyed the art of jade carving in America, and soon after the Conquest, the jade sources were forgotten.

The Maoris of New Zealand used nephrite, starting in about 1000 CE, first as tools and weapons and later for amulets and decorations.

Legends and Lore

Since the beginning of their history, the Chinese have esteemed nephrite more than any other gemstone. From Neolithic times until the beginning of the twentieth century, carved *pi* or *bi* (flat discs with a central hole) were used to worship heaven.

According to a 1596 Chinese encyclopedia, drinking a mixture of jade, rice, and dew water would strengthen the muscles, harden the bones, calm the mind, enrich the flesh, and purify the blood. One who took it long enough could endure heat, cold, hunger, and thirst.

Jade was equally important after death. An elaborate burial shroud made of 2,156 jade tablets sewn together with threads of gold covered the princess Dou Wan (second century BCE). Carved jade amulets were put in the deceased's mouth, and amulets, insignias of rank, and favorite pieces were placed on different parts of the body and clothing. These "tomb jades" were offerings to the gods, but the durable stone was also believed to protect the body from decay.

In the pre-Columbian civilizations of Mexico and Central America, jade had great talismanic power as well. A piece of jade in the mouth of a dead nobleman was believed to serve as a heart in the afterlife. Powdered and mixed with herbs, jade was used as a treatment for fractured skulls, different fevers, and even reviving the dying.

The Maoris of New Zealand also revered jade as a powerful talisman. Typical are the hei-tiki pendants, which are stylized human faces or forms of carved jade that were passed down to the male heirs from generation to generation.

OPPOSITE: Kunz Ax, Olmec style (1200–400 BCE), Oaxaca, Mexico; fashioned from jadeite jade; 27.9 cm (11 in) in height and weighing 7 kg (15.5 lb.).

RIGHT: A Maori jade tiki ornament, New Zealand, of an unknown date.

Occurrences

Nephrite is a metamorphic rock that results from chemical reactions between serpentinite (serpentine rock), adjacent or embedded rock, such as granite, and a hydrothermal fluid; or between granite and dolomitic (magnesian) marble. Jadeite rock forms from fluids squeezed out of subduction zones, those tectonic environments where Earth's oceanic crust dives into the mantle where plates collide. Jadeite jade (*jadeitite* in rock terms) is very rare because its formation also requires special conditions of uplift to exhume the deeply buried (from high pressure) rock. The durability of the jades results in their survival as stream cobbles and boulders, typically the first finds of a deposit.

Canada's British Columbia has been the world's major supplier of nephrite, but the resource is depleted, particularly considering the harsh environment. Here, grayish-green to emerald-green nephrite is found, mostly along the Fraser River. Taiwan and China have been importing most of the British Columbian jade. Siberian nephrite has ascended in the marketplace in recent years, from deposits in West Sayan, west of Irkutsk, and near the north end of Lake Baikal.

Alaska, California, and Wyoming are also sources of nephrite. Jade Mountain in Alaska was located in 1886. The deposits are huge, but remoteness and Arctic conditions limit exploitation. South-central Wyoming has the best-quality nephrite in the Western Hemisphere, but the supply is now severely depleted.

In New Zealand, nephrite is found on South Island, both in situ and as pebbles and boulders in the streams, and exploitation is limited to the indigenous people. The town of Hokitika is the jade center of New Zealand.

Eastern Turkestan (now Xinjiang Province in China) has been the main source of Chinese nephrite since Western trade developed around the time of the Shang Dynasty (ca. 1600 BCE). For millennia, Hotan (Khotan) was the nephrite capital near the sources in the Kunlun-Altai Mountains, which extend for nearly 2,000 km (1,243 mi.) along the edge of the Tarim Basin. These nephrites are mainly white

ABOVE: Nephrite pi disk measuring 31 cm (12¼ in) across from the Ming Dynasty (1368–1644). The pi is the symbol of heaven and one of the most important ritual jades. At burials, it was placed under the body of the deceased.

OPPOSITE: The largest nephrite boulder from Europe ever recorded weighs 2,144 kg (4,727 lb.) and has been polished on one side by Tiffany & Co. George F. Kunz, en route to Russia in 1899, heard that there was nephrite at Jordansmühl in Silesia (now Jordanów Slaski, Poland). He arrived at the quarry at 6 a.m. and breakfasted with the owner, who provided him with a cart and workers. Kunz found the piece, and the quarry owner gave it to him as the discoverer's right. (On loan from the Metropolitan Museum of Art.)

xor near black. Ancient Chinese green nephrite appears to have come from Manasi, in the Tian Shan Mountains. Other occurrences of nephrite are in Taiwan, Australia, Poland, and India.

Myanmar is the main commercial source of jadeite and the only significant source of Imperial jade. It may be one of the oldest sources as well, for prehistoric jadeite instruments have been found in the Mogok region, possibly fashioned from pebbles and boulders recovered from the Uru River and other streams. Jadeite in situ was not discovered until the 1870s. The mines are mostly government-commercial joint ventures, but continued hostilities between the ethnic Kachin people and the government have made the region unsafe (ca. 2014). Exports in 2014, officially through the Myanmar Gems Emporium, are estimated in billions of U.S. dollars, much of that being in jadeite jade, but untold amounts go directly, under the radar, to Yunnan Province in China or through Thailand to other markets.

Jadeite is produced commercially in central Guatemala along the Motagua River and is largely used to feed the tourist market for replicas of Maya jade. Russia holds two deposits and Kazakhstan one, all of which are potentially commercial except for difficult locations for extraction and quality not sufficiently competitive with Myanmar. Jadeitite is found in San Benito County in California and at three localities in Japan. Recently discovered sources in Cuba and the Dominican Republic may have supplied jade to eastern Caribbean peoples from ca. 470 CE until European contact.

Evaluation

The differences in quality and prices of jade are great. Color and translucency are the major considerations in evaluating both nephrite and jadeite. Most of the jadeite jade mined is used as so-called utility grade for making bathroom tile.

The rarest and most valued color for jadeite is pure, even, and intense emerald green. When this color is combined with maximum translucency and smooth, uniform texture, the stone—known as Imperial jade or *fei tsui*—commands an extremely high price. Next in value is lavender. A jade piece has a high value if the color is pure, intense, and uniform, even if it is almost opaque.

Sometimes white jadeite jade is dyed green or lavender. The dyes are not always permanent, and the green frequently fades. Impregnation of semiopaque jadeite jade with resins can produce highly transparent material known as "B jade" (for Grade B) in the marketplace.

As a gemstone, jadeite commands substantially higher prices than nephrite; however, natural white (mutton-fat) nephrite pebbles are highly valued (and now highly priced) in China. Design, craftsmanship, and antiquity are the major considerations in evaluating carvings.

JADE SUBSTITUTES AND THEIR TRADE NAMES

Jade is imitated with mounted jade triplets, glass, and plastic. Many jade substitutes are on the market and many other carving materials are readily confused with jade. Serpentine is probably the most common substitute.

Bowenite (gem serpentine): Korean or immature jade

Amazonite feldspar: Amazon and Colorado jade

Varieties of green grossular garnet: Transvaal or Putao jade

Aventurine quartz: Indian jade

Mixture of idocrase and grossular: American jade or Californite

Soapstone: Fukien, Manchurian, or Hunan jade

Green jasper: Swiss or Oregon jade

Green-dyed calcite: Mexican jade

Chrysoprase: Australian jade

OPPOSITE: Serpentine vase carved in China from Mongolian material, 16.7 cm (6⁹⁄₁₆ in). Serpentine is commonly mistaken for jade.

QUARTZ

Quartz is a very common mineral and, when observed as well-formed transparent crystals, is easily recognized. It comes in a number of colored varieties—amethyst, citrine, rose quartz—but the colorless variety epitomizes the popular concept of crystal. The beauty and symmetry of the pointed "hexagonal" crystals and their water-clear transparency captivate the eye. It is no wonder that this natural gem has had great significance in many cultures throughout human history. Quartz crystals are among the earliest talismans; beads and seals were the first crystalline objects to be fashioned, and "gazing balls" with mystic significance are virtually synonymous with rock crystal. Whether as a crystal gemstone or in polished form, quartz can be found in the earliest prehistoric grave or the most modern collector's cabinet. The New Age attention to transcendental perceptions about quartz revived an ancient tradition.

QUARTZ DATA

Silicon oxide or silica: SiO_2

Crystal symmetry: Trigonal

Cleavage: None

Hardness: 7

Specific gravity: 2.65

R.I: 1.544–1.553 (moderate)

Dispersion: Low

OPPOSITE: Found in McEarl Mine, Hot Springs, Arizona, this rock crystal measures 13.5 cm (5⅓ in) in height.

Properties

The widespread availability (and thus moderate cost) of large, clear pieces in an array of colors provides quartz's appeal as a gemstone; otherwise, it has low brilliance and fire. Quartz is a remarkably pure mineral, but its coloration does require chemical impurities, although only a little—less than one impurity per thousand silicon atoms. Also, irradiation by natural or artificial means is necessary to produce both amethyst and smoky to black quartz; a large amount of the available smoky quartz is artificially irradiated rock crystal—the colorless transparent variety. Similarly, citrine, which is rare in nature, is commercially created by heat-treating natural amethyst.

Quartz has a strong framework crystal structure that makes it hard and free from cleavage—a durable material. It is also a common component of dust, which is the abrasive enemy of all gemstones; this is why quartz's hardness is considered the division between soft and hard gems—those softer or harder than quartz.

Inclusions in quartz are responsible for some very interesting varieties with banded or spangly reflections or just color. Fibers are found in rutilated quartz, cat's eye, and sagenite. In hawk's eye, the blue color comes from blue asbestos intergrowths with quartz that grew together across rock fractures to form veins; if the surfaces of the asbestos fibers oxidize (from water infiltration), iron oxides so produced impart the bronze color of tiger's eye. Small particles, fractures, and fluids are responsible for aventurine, iris quartz, and milky quartz.

Quartz crystals are usually elongate hexagonal-looking prisms capped by a "hexagonal pyramid." However, the crystals only have threefold symmetry. This fact is well demonstrated by the three-bladed pinwheel effect in some amethysts.

Quartz, alone with tourmaline among gemstones, lacks a center of symmetry in its crystal structure; this condition makes it piezoelectric. When pressure is applied across opposing prism faces, they develop opposite charges; relaxation reverses the effect. The property has important application in electronics, but there is no substantiated scientific evidence that humans can directly sense or activate electronic vibrations in quartz.

GEMSTONE QUARTZ VARIETIES, COLORS, AND COLOR SOURCES

Rock crystal: Colorless and transparent

Amethyst: Purple—iron more than aluminum and irradiation

Citrine: Yellow to amber—iron

Morion: Black—aluminum more than iron + irradiation

Smoky quartz or cairngorm: Smoky gray to brown—aluminum + iron and irradiation

Rose quartz: Translucent pink—inclusions of fibers similar to dumortierite

Green quartz, or praziolite: Green—iron + heating

LEFT: "Pinwheel" amethyst weighing 41.17 cts., from Brazil. It shows quartz's trigonal symmetry.

OPPOSITE: Quartz varieties including rock crystal, smoky quartz, citrine, amethyst, rose quartz, and green quartz. Weights range from 13.16 to 489.85 cts. The gems come from various localities.

Historic Notes

The word *quartz* derives from the Slavic *kwardy*, meaning "hard." The Latinized *quarzum* was first recorded in the sixteenth century by the German scholar Georgius Agricola, who made the first scientific classification of minerals. According to him, the term was used by the Bohemian miners of Joachimsthal (now Jáchymov, the Czech Republic).

Rock crystal objects have been found with remains of prehistoric man (75,000 BCE) in France, Switzerland, and Spain. Cylinder seals of rock crystal appeared in the Near East by the fourth millennium BCE. As an amulet and a decorative stone, it found use in Egypt before 3100 BCE. The ancient Greeks and Romans used it extensively in jewelry, valuing its flawless transparency. The Greeks believed that the gemstone was water frozen by the gods to remain forever ice; our term *crystal* derives from the Greek *krystallos*, meaning "ice."

During the Middle Ages and Renaissance in Europe, rock crystal vessels were carved for royal and ecclesiastical use, and for centuries the Japanese and Chinese carved the material, as did the Mayas, pre-Aztecs, Aztecs, and Incas on the other side of the world.

Amethyst was used as a decorative stone before 25,000 BCE in France and has been found with the remains of Neolithic people in different parts of Europe. Before 3100 BCE in Egypt, beads, amulets, and seals were made of this gemstone, and it was highly valued in ancient Greek and Roman societies. An amethyst was the ninth stone in the breastplate of the high priest of Israel and one of the ten stones on which the names of the tribes of Israel were engraved.

In medieval times, amethyst graced royal crowns and bishops' rings. A huge round amethyst adorns the British royal scepter, set for the coronation of James II (1633–1701), and another remarkable amethyst surmounts the sovereign

Amethyst scepter crystal 4 cm (1½ in) across perched on milky quartz from Hopkinton, Rhode Island.

orb. Brazilian amethysts appeared on the European market in 1727 and became highly fashionable and expensive. Amethyst was very popular during the eighteenth century in France and England. A necklace of amethysts was purchased at a very high price for Queen Charlotte (1744–1818), wife of George III of England. Soon after, the price declined as amethysts from Ural Mountains deposits (discovered in 1799) and Brazil increased the supply.

Smoky quartz was used by the Sumerians in the valley of the Euphrates and by the Egyptians before 3100 BCE. Beads surviving from Roman times are fairly common. The stone was popular with the Navajos and other Native Americans. Smoky quartz from the Cairngorms, a mountain range in Scotland, is also called *cairngorm*; the stones have been set into brooches and the handles of dirks.

Rose quartz, named for its pink to rose color, was first used by the Assyrians around 800–600 BCE and later by the Romans, but it was not common in the jewelry of the ancient world. Today it is very popular for bead necklaces, small carvings, and cabochons.

Citrine was considered a gem during the Hellenistic Age in Greece (323–330 BCE) and was moderately popular for intaglios and cabochon ring stones through the first and second centuries CE in Greece and Rome. It has had ongoing use in jewelry but has never achieved the prominence of amethyst. The name derives from the French *citron*, meaning "lemon," an allusion to its color. Agricola applied the term to the yellow variety of quartz in the sixteenth century.

TOP: Amethysts of 163.50 cts. and 88.20 cts., both from the Ural Mountains in Russia.

ABOVE: An amethyst group measuring 7.5 cm (3 in) across, from Thunder Bay, Ontario, Canada.

OPPOSITE: A Fu dog carving from China, rose quartz, 15 cm (6 in) high.

ABOVE: A 45.71-ct. ametrine (trystine), from Puerto Suárez, Bolivia, and an amethyst brooch set with diamonds. Ametrine is bicolor amethyst-citrine, discovered in 1977.

Legends and Lore

The Greeks believed that amethyst would prevent intoxication (the Greek word *ame-thystos* means "not intoxicated"), and, in legend, it also calmed anger and relieved frustrated passion. In the sixteenth century, a French poet wrote a myth about the Greek god of wine, Bacchus. To avenge an insult, Bacchus declared that the first person he met would be devoured by his tigers. This person happened to be Amethyst, on her way to worship at the shrine of Diana. As the beasts sprang, Diana turned the girl into a clear crystal. Repenting, Bacchus poured grape juice as a libation over the stone, thus giving the gem its beautiful purple color.

In medieval Europe, amethyst was worn as a soldier's amulet for protection in battle. In addition, amethyst made men shrewd in business matters, according to Camillus Leonardus, a sixteenth-century authority on precious stones.

The allure of rock crystal has been virtually universal. The Japanese regarded it as "the perfect jewel, *tama*": a symbol of purity, the infinity of space, patience, and perseverance. In places as far apart as North America and Burma, rock crystal has been considered a living entity. Native American Cherokees not only used it as a talisman for hunting, they periodically "fed" the stone by rubbing it with deer's blood; the crystal was similarly nourished by the Burmese.

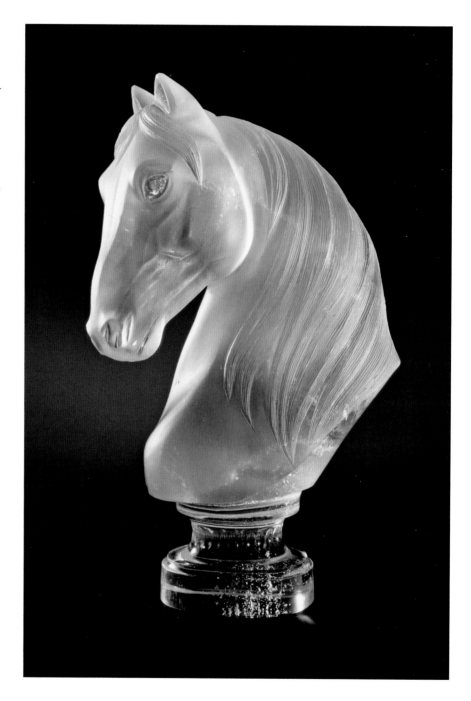

ABOVE: Russian rock crystal carving from the nineteenth century measuring 13.2 cm (5¼ in).

Crystal balls that the Crusaders brought from the Near East were reputed to possess magical powers. Until the late 1800s, rock-crystal balls set in metal known as charmstones were employed to cure or protect against diseases of cattle in Ireland and Scotland. In modern times, quartz crystal has attained enormous popularity from the belief that it is a psychic amplifier with healing attributes.

This illustration depicts the ancient Stone of Ardvorloch, a Scottish rock-crystal charm stone measuring 2.5 cm (1 in) and mounted in silver. From *Archaeological Essays* by Sir James Y. Simpson, 1872.

cavities in the Paraná basalts are numerous and up to several meters in height. Amethyst is found also in Canada, Russia, Zambia, and in the United States in Arizona, North Carolina, and Georgia. The major commercial source of natural citrine is Brazil (Minas Gerais, Goiás, Espírito Santo, and Bahia) and Uruguay.

Evaluation

Among the many varieties of quartz, amethyst is the most expensive. Intense and uniform purple is the most desired. Any flaws diminish the price considerably. Amethyst may be confused with purple sapphire or spinel. It can be imitated with synthetic sapphire or glass. Synthetic amethyst is manufactured in Russia.

Clarity is the factor in evaluating rock crystal. The price is moderate, except for very large flawless pieces. Herkimer diamond, Arkansas diamond, Arizona diamond, Cape May diamond, Alaska diamond, and Cornish diamond are some of the misnomers used for rock crystal. Synthetic rock crystal is used in industry rather than in jewelry. Poorly formed or coated quartz crystals are polished and sold as natural crystals or points. This practice is misleading; the quartz is natural, but the faces are not.

For rose quartz, the deeper the color and the more transparent the gem, the greater its value. Rose quartz showing chatoyancy is cut into spheres, for which the intensity and sharpness of the star are the keys to value. Rose

Occurrences

Quartz forms in a wide variety of environments, but gemstone crystals usually require the openings in rocks, such as veins, cavities, and pockets, to grow to perfection and adequate size. Crystal-lined pockets, called geodes, are a familiar source of quartz. In pegmatites containing quartz and sufficient radioactive minerals to provide the necessary irradiation, amethyst and smoky quartz will develop.

There are many commercial sources of gem-quality rock crystal. Brazil is the principal source for all varieties of quartz. Arkansas near Hot Springs is important for rock crystal in the United States. The most prolific producers of amethyst are Brazil and Uruguay, where crystal-lined quartz is sometimes dyed, but the dye fades. The value of both citrine and smoky quartz is determined by clarity and the attractiveness of their colors. Quartz of yellow and golden brown to orange brown is rare. Citrine is often sold fraudulently as topaz, whose color is richer. However, due to its lack of cleavage, citrine is tougher, wears better, and is less expensive.

INCLUSIONS IN QUARTZES

Milky quartz: white—fluids, mainly water

Aventurine: Green or brick red—chromian mica or hematite flakes

Rutilated quartz: Golden reflecting—rutile needles

Iris quartz: Iridescence—numerous small cracks

Sagenite (or Venus hair, Thetis hair): A netlike pattern of needles—rutile, black tourmaline, green actinolite, or epidote

Cat's eye: Chatoyancy in several color varieties—fibers of rutile

Tiger's eye: Bronze chatoyancy—brown iron oxides on asbestos fibers

Hawk's eye: Blue chatoyancy—blue asbestos

ABOVE: The rock crystal statue of Atlas holding up the world is 12 cm (4⅝ in) in height. It was carved during the nineteenth century in Russia from a crystal found in the Ural Mountains.

OPPOSITE: Group of three intergrown rutilated quartz crystals measuring 6.7 cm (2⅝ in) in height, from Itabira, Minas Gerais, Brazil.

CHALCEDONY & JASPER

These gemstones were as highly prized by our earliest ancestors as they are by today's lapidary hobbyists. The basis of their appeal comes from the literally hundreds of colorful varieties that can be found. Both are made up of submicroscopic quartz grains—thus are varieties of quartz (see pages 128–39)—and owe their bonanza of colors and patterns to included minute grains of other pigmenting minerals. Chalcedony differs from jasper in that its tiny crystals are parallel fibers rather than sugar-like grains. Distinguishing them requires a microscope, although typically chalcedony is banded and translucent. Listing all the varieties of these gemstones would be daunting to a philologist, let alone a mineralogist. Only the most important and well-known varieties of chalcedony and jasper are discussed here.

OPPOSITE: Carnelian vase carved in China; it measures 10.5 cm (4⅛ in) across.

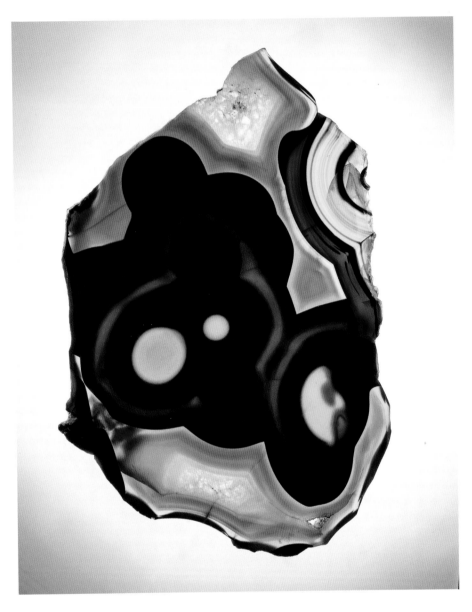

of crystals, such as symmetry, are not visible to the naked eye.

In chalcedony the tiny quartz fibers form layers in a velvetlike pile. In the 1990s, it was recognized that some of the silica in some chalcedonies is a different mineral, moganite, with slightly more crystal symmetry but very similar to quartz. The layers stack up one upon the other, often producing a banded appearance as seen in the best-known chalcedony variety, agate. The fibrous structure imparts substantial toughness, too. Layers can be translucent to opaque and vary from gray to white (when they are free of impurities) to almost any color (when they are pigmented with an appropriate impurity). Except for white layers, porosity is pronounced; the gemstones are easily dyed. Onyx, a black-and-white variety, is naturally rare but is commercially produced by soaking pale agate in a sugar solution and then carbonizing the sugar in sulfuric acid, rendering the gemstone black and white.

Jasper's granular texture makes it tough and generally more opaque than chalcedony, and jasper lacks the other's banding. While commonly red to ochre from iron oxide pigments, jasper can occur in a multitude of colors. Some materials have mixed textures of both chalcedony and jasper juxtaposed in anywhere from millimeter- to centimeter-scale patches. Gemstone varieties that show either or both textures include bloodstone and chrysoprase.

Properties

The gemstone properties of chalcedony and jasper are the properties of quartz—good hardness and durability. As essentially superfine-grained rocks, the directional properties

The layered texture of agates, particularly onyx and sardonyx, has made them very popular materials for carving intaglios and cameos. Cameos are usually carved with the white layer in relief and the colored layer as background. In intaglios, the figure is incised through the dark layer to reveal the white layer—or the reverse.

OPPOSITE: A polished agate slab from an unknown locality measures 18 cm (7 in) across.

ABOVE: Agate cameo, its rough from Uruguay, measuring 4.7 cm (1⅞ in) in length.

RIGHT: This onyx clock face, carved cameo style in Germany, displays carved female figures. The diameter is 11.5 cm (4½ in).

CHALCEDONY VARIETIES

Agate: All forms with parallel to concentric banding, transparent to opaque.

Bull's eye agate: Bands form concentric circles.

Iris or fire agate: Iridescent from alternating thin layers of iron oxide and chalcedony.

Onyx: Bands are black and white—popularly misconstrued to be all black.

Sardonyx: Bands are brown to ochre and white.

Bloodstone or heliotrope: Plasma with red hematite or jasper spots and blotches.

Carnelian: Translucent red brown to brick red from hematite.

Chrysoprase: Translucent apple green from nickel serpentine.

Chrysocolla chalcedony: Stained blue-green by microscopic grains of chrysocolla (a copper silicate) and perhaps other green copper minerals.

Moss agate: Translucent light-colored body with black, brown, or green moss-looking to branchlike (dendritic) inclusions, usually dark oxides. "Mocha stone" is moss agate from a source near Mocha in Yemen.

Plasma: Opaque leek to dark green from various green silicate minerals.

Prase: Translucent leek green from chlorite inclusions.

Sard: Translucent light to chestnut brown from iron oxides and hydroxides.

OPPOSITE: Moss agate from India. The largest is 7.5 cm (3 in) in diameter

ABOVE: An assortment of jaspers, chalcedonies, and other ornamental stones, including a heliotrope (bloodstone—green and red cylinder); a banded brown, white, and pink jasper cabochon; a carnelian pendant 6 cm (2⅜ in) high; a faceted blue chalcedony; an obsidian bowl; an onyx intaglio; and sodalite, prehnite, turquoise, and jadeite cabochons.

Historic Notes

The term *chalcedony* may derive from the ancient Greek port of Chalcedon. The terms *chrysoprase* and *prase* come from the Greek *chrysos* and *prase*, meaning "golden" and "leek."

Carnelian derives from the Latin *cornum*, meaning "cornel berry" or "cornelian cherry." *Heliotrope* (bloodstone) derives from the Greek *helio*, meaning "sun," and *trepein*, meaning "turning." *Jasper* derives from the Greek *iaspis*, of Oriental origin but unknown significance.

Sard comes from the Greek Sardis, capital of Lydia in Asia Minor. *Agate* is named for the Achates (Dirillo) River in Sicily, a major source of the gem, according to Theophrastus. *Plasma* alone is a use-derived name; the Greek word from which it comes means "something molded" or "something imitated."

The oldest jasper adornments date back to the Paleolithic period. Agate has been found with the remains of Stone Age man in France (20,000–16,000 BCE); and agate, carnelian, and chrysoprase were used by the Egyptians before 3000 BCE. Magnificent agate and jasper jewelry has been found in Harappa, one of the oldest centers of the Indus civilization. Sard was used by the Mycenaeans (1450–1100 BCE) and the Assyrians (1400–600 BCE). Carnelian and sard were favorite stones of Roman gem engravers. Carnelian seals have been esteemed by the Muslims; the prophet Muhammad was said to have worn one himself. Prase was used as a gemstone in Greece around 400 BCE. Mining of agate at aboout the same time in India has been documented, although the gem was probably used much earlier.

The small German towns of Idar and Oberstein were a source of agate, jasper, and other stones in Roman times, and during the fifteenth century, an agate industry was established there. It flourished until early in the nineteenth century, when the mines were depleted, and many skilled miners and lapidaries went elsewhere. In 1827, German settlers discovered rich chalcedony deposits in Brazil and Uruguay. By 1834, Brazilian agate was being exported to Germany. Although Idar-Oberstein is no longer a supply source, it is renowned for the quality and artistry of its gem craft. Currently, it imports raw material from about 100 countries and employs more than 500 gem polishers and numerous engravers and wholesale gem dealers.

OPPOSITE: Carnelian in an Islamic necklace with tassels; a Chinese belt buckle measuring 6 cm (2⅜ in) across; seventh-century Merovingian necklace spanning 28 cm (11 in), found in France near Soissons; and Native American bracelets.

Legends and Lore

As ancient gems, the chalcedonies and jaspers have accrued the lore of the ages. Bloodstone preserves an owner's health and protects him or her from deception (Damigeron, first century CE). Sard has medicinal virtue for wounds (Epiphanius, fourth-century bishop of Salamis in Cyprus) and protects the possessor from incantations and sorcery (Marbode in the eleventh century). Chrysoprase strengthens the eyesight and relieves internal pain (eleventh-century Byzantine manuscript of Michael Psellus). Carnelian gives an owner courage in battle (Ibn al-Baitar, botanist of the thirteenth century) and helps timid speakers become both eloquent and bold.

Perhaps the most intriguing virtue of all is noted in Volmar's thirteenth-century *Steinbüch*: a thief, sentenced to death, may escape his executioners immediately—if he puts chrysoprase in his mouth.

Occurrences

Chalcedony and jasper are geologically common, formed in cavities, cracks, and by replacement where low-temperature silica-rich waters percolate through sediments and rocks, particularly those of volcanic origin. Chalcedonies are common the world around. Brazil, Uruguay, and India produce all varieties of chalcedony and jaspers. With the exception of sard and plasma, all chalcedonies come from diverse localities in the United States.

Additional sources are: chrysoprase—Australia, Poland, Germany, Tanzania, Zimbabwe, and Russia; carnelian—South Africa, Russia, and China; agate—Mexico, Namibia, and Madagascar; jasper—Venezuela, Germany, and Russia; bloodstone—Australia and China.

Evaluation

The attractiveness of colors and patterns determines the value of all varieties. The naturally colored stones have higher prices than the artificially colored. Chrysoprase is rare and the most valuable variety. Translucency is an important consideration for chrysoprase, carnelian, sard, agate, and prase. (Prase is presently rarely used in jewelry, however.)

ABOVE: *Pas de Danse*, carved by Georges Tonnelier (ca. 1900), stands 21.5 cm (8½ in) tall. Its chalcedony rough is from Uruguay.

OPPOSITE: A chalcedony vase carved in China measures 10.6 cm (4¼ in) high.

GARNET

Garnets are not just red; they come in all colors except blues. The range of garnet colors may be surprising to many, as well as the fact that new varieties of gem garnet were discovered as recently as the 1970s. Tsavorite was unearthed in 1968 near Kenya's Tsavo National Park and named for it by Tiffany promoters. It is a beautiful gemstone that rivals emerald but has been in short supply. And around 1970 in eastern Africa, a reddish orange garnet was found during the search for the purplish pink rhodolite garnet that was particularly desired in Japan. Attempts to sell the Japanese on the new garnet were futile, so the stone was called *malaia*, a Swahili word meaning "outcast" and "prostitute." To the Africans' surprise, in the late 1970s Americans found the maligned garnet very attractive. But the name has stuck.

GARNET DATA

Garnet is the name of a group of silicate minerals; the gemstone garnets can be described in terms of five limiting members. There's extensive solid solution between minerals listed within each grouping but not between the groups.

Pyrope: $Mg_3Al_2(SiO_4)_3$
Almandine: $Fe_3Al_2(SiO_4)_3$
Spessartine: $Mn_3Al_2(SiO_4)_3$

Grossular: $Ca_3Al_3(SiO_4)_3$
Andradite: $Ca_3Fe_2(SiO_4)_3$

Crystal Symmetry: Cubic
Cleavage: None
Hardness: 6.5–7.5
Specific gravity: 3.5–4.3
R.I: 1.714–1.895 (moderate to high)
Dispersion: Moderate

OPPOSITE: Etched mass of spessartine garnet sprinkled with pyrite and measuring 6 cm (2⅜ in.) across and a 98.61-ct. cut stone from Amelia Court House, Virginia.

Properties

Garnets are colorful, lively, and durable—a fine gemstone group, but complex. There are many varieties and many mineral species—like a fruit market with Granny Smith and Delicious apples as well as Concord and green grapes. The colors of the gemstone garnets vary with species as well as with minor substitutions of transition metals into the structure. The iron- and manganese-bearing garnets are intrinsically colored (idiochromatic), whereas those without transition elements are colorless in the pure form (allochromatic). Grossular has the greatest range of colors, and andradite has the highest brilliance and fire—particularly the superb green variety, demantoid. The mistaken concept that garnets are red derives from the predominant use of almandine and pyrope as gems.

Garnet, particularly almandine, can develop fibrous inclusions of several possible minerals in three perpendicular directions; such gemstones can show four- or six-rayed stars when fashioned as cabochons. The garnet's cubic symmetry leads to multifaced equidimensional crystals that can vary greatly in size; some small ones look like natural beads.

GEMSTONE GARNET SPECIES, VARIETIES, COLORS, AND CAUSES OF COLOR

SPECIES	VARIETIES	COLORS AND CAUSES
Pyrope		Colorless, pink to red from iron
	Chrome pyrope	Intense red from chromium
Almandine		Orangey red to purplish red
Pyrope-almandine		Reddish orange to red purple
	Rhodolite	Purplish red to red purple
Spessartine		Yellowish red to red purple
Almandine-spessartine		Reddish yellow to purple
Pyrope-spessartine		Greenish yellow to purple
	Malaia	Yellowish to reddish orange to brown
	Color-change garnet	Blue green in daylight to purple red in incandescent light due to vanadium and chromium
Grossular		Colorless, orange from ferrous iron, also pink, yellow, and brown
	Tsavorite	Green to yellowish green from vanadium
	Hessonite	Yellow orange to red orange from manganese and iron
Andradite		Yellowish green to orangey yellow to black
	Demantoid	Green to yellow green from chromium
	Topazolite	Yellow to orangey yellow

OPPOSITE: Spessartine crystals up to 1.5 cm (⁹⁄₁₆ in) across on quartz from Nangarhar Province, Afghanistan; a 28.41-ct. almandine from Tanzania; and an 8.97-ct. round brilliant-cut pyrope from Macon County, North Carolina.

RIGHT: Rhodolite garnet from Tanzania, weighing 24.51 cts.

FAR RIGHT: A demantoid garnet from Poldenwaja in the Ural Mountains of Russia, weighing 4.94 cts.

Historic Notes

The word *garnet* derives from the Latin *granatum*, meaning "pomegranate," and alludes to the crystal's red color and seedlike form. Red garnet gems date back thousands of years. Excavations of lake dwellers' graves in the Czech Republic have uncovered garnet necklaces and suggest use of the material in the Bronze Age. Other findings indicate widespread use of the gems for beads and inlaid work in Egypt before 3100 BCE, in Sumeria around 2300 BCE, and in Sweden between 2000 and 1000 BCE. Garnets were the favorite stones in Greece in the fourth and third centuries BCE and remained popular during Roman times. Garnet-inlaid jewelry has been discovered in southern Russia in graves of the second century CE. Over four thousand garnets decorate jewelry that was found when a seventh-century ship burial was excavated in East Anglia in 1939. Aztecs and other Native Americans used garnets in their ornaments in pre-Columbian times.

Pyrope garnets were the basis for a thriving jewelry and cutting center in Bohemia (Czech Republic) that started in about 1500. Until the late nineteenth century, the Bohemian deposits were the world's major source of the stone.

LEFT: A hessonite (grossular) engraved with Christ's head, from the Vatican collection. This piece measures 3.6 cm (1⅜ in) in height.

OPPOSITE: Engraved almandine garnet bowl from India with a diameter of 5.5 cm (2⅛ in).

NAMES OF THE GARNETS

Pyrope: From the Greek *pyros,* meaning "fiery" and alluding to the stone's deep red color

Almandine: From Alabanda, an ancient garnet source in Asia Minor (now Turkey)

Rhodolite: Derived from two Greek words that mean "rose stone"

Spessartine: Named for Spessart, the Bavarian district where the gem was first found

Andradite: Named after mineralogist J. B. d'Andrada, who described a variety in 1800

Topazolite: Similar to topaz in color

Demantoid: Derived from the Dutch *demant*, meaning "diamond," named for its diamond-like brilliance

Grossular: Derived from the botanical name of the gooseberry, *R. grossularia*, alluding to similarity between the colors of the berries and some pale green grossulars.

Legends and Lore

A single large garnet provided the only light on Noah's ark, according to the Talmud. During the Middle Ages, garnet was regarded as a gem of faith, truth, and constancy. As late as 1609, Anselmus de Boodt contended that garnet drives away melancholy.

Like other red stones, garnet was considered a remedy for hemorrhage and inflammatory diseases and a general protection against wounds, a belief that has been revived among some New Age adherents. In contrast, some Asiatic tribes believed that garnet bullets would be more deadly than those of lead. Accordingly, in 1892, the Hunzas of what is now northern Pakistan used garnet bullets (some of which have been preserved) against British troops during hostilities on the Kashmir frontier.

Occurrences

Garnets form in many metamorphic and some igneous rocks. Almandine is the common metamorphic garnet; spessartine is similar, but the purest orangey ones form in pegmatites. Pyrope crystallizes at high pressures. Grossular and andradite form in contact metamorphic zones, particularly next to marble.

Superb rubylike pyropes are found in diamond-bearing kimberlites in South Africa and Russia. Fine-quality but small pyropes are found in Arizona, New Mexico, and Utah. Other sources are Kenya, Mozambique, Tanzania, Australia, Brazil, and Myanmar.

Major sources of almandine are India, Sri Lanka, and Brazil. Star stones are found in Idaho and India.

Rhodolite, more transparent than pyrope and almandine, was originally found in Lower Creek, North Carolina, in 1882. The major commercial source is Tanzania; rhodolite is also found in India, Sri Lanka, Zimbabwe, and Madagascar.

Gemstone spessartine's rare occurrences include Brazil; Tanzania; Zambia; Ramona, California; and Amelia Court House, Virginia, the major source in the late nineteenth century.

Malaia, the new variety, comes from the Umba Valley in Tanzania. Color-change pyrope-spessartine is also found in eastern Africa. Demantoid was first found in about 1851 in placers in the Ural Mountains. A newer deposit is in Chukotka in Russia. Other occurrences are in Madagascar; Namibia; Zaire; Korea; Iran; and Val Malenco, Italy.

OPPOSITE: Spessartine crystal (modified dodecahedron) 1.5 cm (⁹⁄₁₆ in) across on smoky quartz. The specimen was found in Ramona, California.

ABOVE: Tsavorite gem gravel and an 8.16-ct. stone, all from eastern Africa, probably Taita Hills, Kenya.

Kenya and Tanzania are the only sources of tsavorite. Other sources of gem grossular are Sri Lanka (yellow, brown, pink, red); Asbestos, Canada (yellow, brown to pinkish); and Chihuahua, Mexico (large crystals, seldom transparent). Green, compact, fine-grained grossular containing small black specks resembles jade and frequently is sold for it. Examples are Transvaal jade, found near Pretoria in South Africa, Putao jade from northern Myanmar, and material from the Yukon Territory in Canada and California. Massive white grossular from Myanmar is used for carvings and often sold as jade.

Evaluation

Purity of color, clarity, and size are the most important considerations. Green garnets are the most highly prized, but the market is plagued by poor availability. Demantoid is the most valuable of the garnets; among all gems, it is prized for its beauty and rarity. Emerald green, transparent, flawless stones are extremely valuable. Brownish red garnets are less valuable than the pure red. The price per carat for a fine-quality garnet increases with size.

GEMSTONES CONFUSED WITH GARNETS

Pyrope and almandine: Red spinel and ruby. (Pyrope is occasionally marketed as Arizona ruby, Cape ruby, Elie ruby, and Fashoda ruby; all are misleading names.)

Rhodolite: Plum-colored sapphire and tourmaline

Grossular: Emerald, topaz, zircon, and jade

Demantoid: Green diamond and zircon

Spessartine: Zircon and grossular garnet

157

PEARLS &
OTHER ORGANIC GEMS

Pearl, amber, coral, and jet share an organic origin and, in that sense, form a gem group.

Pearl

Pearls are finished gems when found. Their beauty has been valued for centuries, but their popularity has varied over time. In 1916, millionaire Morton Plant wanted to purchase a magnificent rope of pearls at Cartier's for his wife. The price was $1 million! He proposed an exchange—a piece of real estate for the necklace—and Cartier agreed. In 1956, these magnificent pearls were auctioned at Parke-Bernet and brought only $151,000.

The real estate is the Fifth Avenue landmark building that Cartier still occupies, one of the most valuable corners in New York City.

PEARL DATA

Pearls are built up from layers of nacre.

Nacre composition: Aragonite, $CaCO_3$ (82–86%); conchiolin—a horny organic substance (10–14%); and water, H_2O (2%)

Luster: Pearly

Cleavage: None; pearls are very tough

Hardness: 2.5–4.5.

OPPOSITE: Manchu headdress pendant with Imperial jadeite, pearls, sapphires, and pink tourmalines.

Properties

Lustrous pearls are produced by mollusks having a nacreous (mother of pearl) lining in response to foreign irritants such as parasites or sand grains. Layers upon layers of nacre are deposited on the object, forming a pearl's onion-like structure. Conchiolin (a horny substance) binds aragonite microcrystals together around the inclusion. The crystals overlap, producing a slightly irregular surface that feels rough when rubbed across the teeth, a reliable way of distinguishing natural and cultured pearls from imitations. The luster is caused by the scattering and interference of light by the concentric layers. The light interference and diffraction produce an overtone color called "orient." Pearls are semitransparent to opaque and are divided into three color groups: white, black, and colored.

Pearl-bearing mollusks inhabit both salt and fresh water. Saltwater pearls are the more highly prized for jewelry. They come principally from three species of oysters of the *Pinctada* genus. The major source of freshwater pearls are mussels of the genera *Hyropsis* and *Cristaria*.

Cultured saltwater pearls are produced by artificial stimulation of the natural process. Baby oysters ("spat") are cultured in plastic cages in protected waters. After three years, a bead of mother of pearl and a small piece of oyster mantle tissue are inserted into the body of each oyster. Nacre is secreted around the bead by the foreign tissue. The oysters are returned to the culture cages in the sea. After three years, the oysters are recovered and the pearls removed. The largest Japanese seawater cultured pearls have a diameter of about two-fifths of an inch (10 mm). Cultured freshwater pearls only involve tissue implants.

Freshwater pearls from the United States, and a shell measuring 9.5 cm (3¾ in) wide and containing a blister pearl and a long hinge pearl.

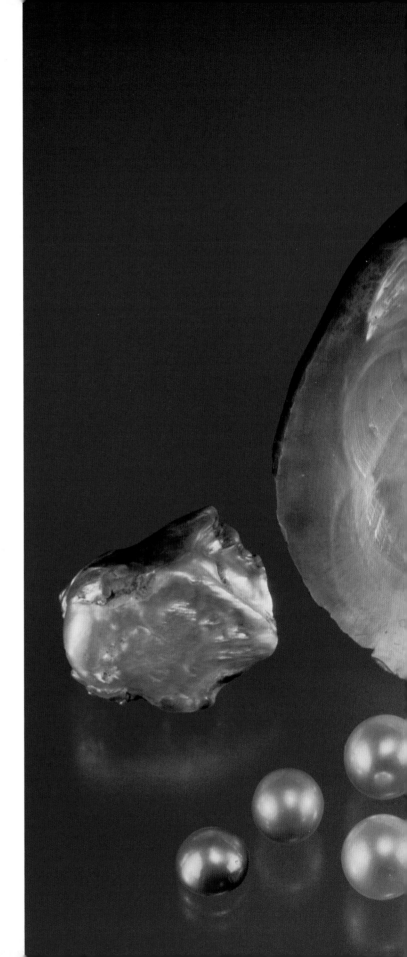

Acids, including those from the human skin, are very damaging to pearls, as are cosmetics, perfume, and hair sprays. Excessive dryness or humidity in the air shortens the life of a pearl. Since pearls are soft, jumbling them together in a box with other gems that scratch them is most destructive. Pearl weight is measured in grains rather than carats: 1 grain = 0.25 carat.

PEARL COLORS AND SHAPES

COLOR: The sum of body color plus overtone color or orient

White: White—white body color, no overtone

Cream—cream body color, no overtone

Light rose—white body color, pinkish overtone

Cream rose—cream body color, deep rose overtone

Fancy—cream body color, rose and blue overtones

Black: Black, gray, bronze, dark blue, blue green, green body colors with or without metallic overtones

Colored: Red, purple, yellow, violet, blue, or green body color—more common in freshwater pearls

SHAPES

Round, pear, drop, egg, button (with flat back), mabe (composite spherical shapes), baroque (irregularly shaped), blister (attached to shell), seed (unsymmetrical, less than 0.25 grain)

Historic Notes

By 2200 BCE, pearls were tax and tribute in China; a Chinese dictionary written before 1000 BCE specifies pearl as a product of western provinces. Embroidered on costumes and made into ropes and ornaments in ancient Persia (Iran), pearls were the privilege of royalty. In Rome, they were the most admired gems and so frequently worn that the philosopher Seneca (ca. 54 BCE–39 CE) criticized women for wearing too many of them.

Throughout the Middle Ages in Europe, pearls were royal gems exclusively, although the Crusaders brought scores of them from the Near East. When Catherine de' Medici (1519–89) came to France to marry Henry, Duke of Orleans, in 1532, she brought six strings of fine pearls and twenty-five large single pearls, which she later presented to Mary

Dowager Empress Tz'u-hsi (center, seated), ca. 1890, wearing her pearl cape, with ladies of the royal court.

Stuart (1542–87). After Mary's death, the Medici pearls were purchased for a trifle by Elizabeth I (1533–1603). In all her portraits, Elizabeth is accoutered with pearls. Men also wore pearls, as demonstrated by royal portraits of the period. In 1612, an edict of the Duke of Saxony stated that the nobility were not to wear dresses embroidered with pearls, and professors and doctors of the universities and their wives were not to use any pearl jewelry. During the late Renaissance, baroque pearl pendants in the forms of dragons, mermaids, and centaurs were popular in Europe. Pearl's prestige continued into the nineteenth century. The gem collection of Dowager Empress Tz'u-hsi (1835–1908) included thousands of precious gems, many pearls among them, and she owned a cape embroidered with 3,500 pearls, each the size of a canary's egg.

In the early twentieth century, pearl prices remained prohibitive for many, but by the 1920s wealthy American women had ropes of pearls rivaling those of European royalty and Asian potentates. In the 1930s, two events changed pearl's future drastically—the Great Depression and the introduction of cultured pearls.

The spherical cultured pearl was first produced in Japan in about 1907. The creator of the industry is Kokichi Mikimoto (1858–1954), known as "The Pearl King." Fine jewelry stores initially rejected cultured pearls; Tiffany & Co. did not sell them until 1956. Today, although natural pearls may be up to ten times more expensive than equivalent cultured pearls, the natural ones amount to less than 10 percent of the total pearl trade in value.

ABOVE: Edith Kingdon Gould, wife of financier George Jay Gould, photographed wearing ropes of pearls, ca. 1900.

Legends and Lore

Pearls, in Indian mythology, were heavenly dewdrops that fell into the sea and were caught by shellfish under the first rays of the rising sun during a period of full moon, a belief adopted by Europeans. According to Hebrew legend, pearls are the tears shed by Eve when she was banished from Eden. For the ancient Chinese, pearls represented wealth, honor, and longevity. Pearls were widely used as medicine in Europe until the seventeenth century. The lowest-grade pearls are still ground and used as medicine in Asia.

Occurrences

Commercial fishing for natural pearls is essentially nonexistent, but in the past it was important in the Persian Gulf and the Gulf of Mannar between India and Sri Lanka. In addition to Japan, cultured saltwater pearls are produced in Australia and a few equatorial islands in the Pacific, where warmer waters and a larger mollusk permit the growth of larger pearls. The largest and finest pearls, famous for their creamy color and fine pink orient, are produced in Myanmar's pearl farms. South Sea pearls are cultured on the north coast of Australia and around Tahiti.

There were once more than one hundred freshwater pearl farms in Lake Biwa, Honshu, Japan, but pollution led to a collapse of the pearl industry, which never recovered. Biwa pearls are usually white with fine luster and baroque shape. Freshwater cultured pearls are mostly produced in China, although Tennessee still has production.

Evaluation

Distinction between natural and fine cultured pearls can be accomplished conclusively only by X-radiography. Important factors for both are size, shape, color, and orient. The most valued shape is a perfect sphere, followed by symmetrical drop, pear, and button. A perfect pearl is a semitranslucent sphere with even color, fine orient, deep luster, and fine texture. Highly valued colors are white and cream with pink overtones and black with iridescent green orient. Most cultured pearls are bleached. Many are tinted pink and sometimes dyed to imitate naturally colored black pearls.

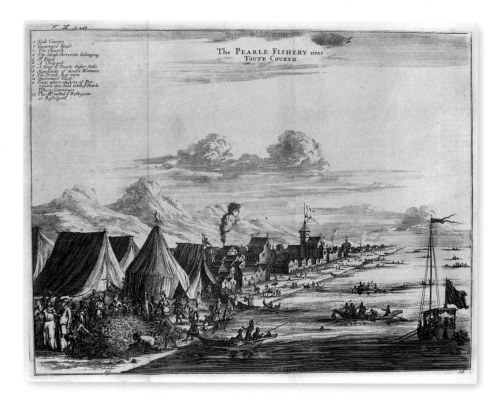

The PEARLE FISHERY near TOUTE COURYN.

Occasionally, pearls are irradiated to produce gray, gray blue, and black colors. The colors are permanent.

Pearls in a necklace should be matched in color, luster, and translucency and strung with knots between them to prevent rubbing—also ensuring that only one pearl will be lost if the string breaks.

The thickness of the nacreous layers is important for cultured pearls. Pearls with a coating of less than one-fiftieth inch are considered low quality. Lacquer is often applied to prevent cracking and wear. Cheap imitation pearls are made of plastic or glass beads with a thin coating of synthetic pearl essence. Finer imitations consist of opalescent glass beads dipped many times into a solution of guanine (manufactured from fish scales), then polished and coated with lacquer to prevent discoloration. Majorcan imitation pearls are known for their good quality.

OPPOSITE: An engraving of a pearl fishery near Toute Couryn (Tuticorin, India, on the Bay of Bengal), 1744. From *A Collection of Voyages and Travels*, by Awnsham Churchill, John Churchill, John Locke, and John Nieuhoff, 1744–46.

ABOVE: Small lead Buddha figures were implanted in a live freshwater mussel and became covered with mother of pearl. The shell measures 11.3 (4⅖ in) across.

RIGHT: A cameo carved into an emperor helmet shell from the West Indies, *Chariot of the Muses*, a late nineteenth-century Italian work.

Amber

Greek philosopher Thales (sixth century BCE) noted that, after it has been rubbed, amber attracts lightweight objects. The Greeks termed the substance *elektron*, a word associated with the sun. Thus the Greek name for amber is the word from which words like *electron* and *electric* derive. Amber derives from the Arabic *'anbar*, meaning "ambergris," a substance obtained from the sperm whale and used in making perfumes.

AMBER DATA

Chemical formula: A mixture of hydrocarbons

Cleavage: None, but sometimes brittle

Hardness: 2–2.5

Specific gravity: 1.05–1.096

R.I.: 1.54

Luster: Resinous

Colors: Yellow, brown, whitish, or red; occasionally green and blue caused by fluorescence or interference of light by included air bubbles

ABOVE: This specimen is an extinct termite, Mastotermes electrodominicus, in Dominican amber, 4.4 cm (1⅓ inches) long.

Properties

Amber is composed of fossilized natural botanic resins of various sorts. Amber is transparent to translucent, often found in sizable pieces, and often contains interesting inclusions—flora and small arthropods trapped by the once-fluid resins. Such fossils date to as early as 230 million years ago, during the Triassic period. Amber is soft but relatively tough, capable of being drilled and carved. Its specific gravity is so low that it floats in a saturated salt solution—a quality that distinguishes it from substitutes, which sink.

Some reserve the term "true amber" for amber from the Baltic region, sometimes called "succinite." Baltic amber is derived from various coniferous trees that lived 30 to 60 million years ago. Dominican amber is somewhat younger than Baltic amber and probably derives from a leguminous plant. Dominican amber is also softer than Baltic amber.

Amber dress ornament carved in China and of Burmese origin, 7 cm (2¾ in) across.

CLASSIFICATION OF BALTIC AMBER

Clear amber: Transparent

Fatty amber: Full of small air bubbles, resembling goose fat

Bastard amber: Clouded because of the presence of many bubbles

Bone amber: White or brown, more opaque than bastard amber

Foamy or frothy amber: Opaque, with a chalky appearance

Historic Notes

Amber pendants, beads, and buttons dating to 3700 BCE have been found in Estonia, and amber treasures found in Egypt date to as early as 2600 BCE. Amber beads from 2000 BCE have been found in Crete and Mycenae, and graduated beads of a similar age in England. In 1000 BCE, the Phoenicians were trading Baltic amber in the Mediterranean region. In Etruria (west-central Italy), amber was used in fashioning inlays, beads, scarabs, and small-figure pendants. Amber has been burned as incense since early Christian times.

During the Middle Ages in Europe, the demand for use as rosary beads consumed the available amber. As the supply increased, so did amber's popularity. The skill of amber carving reached a peak in the sixteenth and seventeenth centuries; examples of carved objects include chalices, candlesticks and chandeliers, religious sculpture, and jewelry. During the nineteenth century, amber jewelry was very popular, although attention focused on the intrinsic value of the gem rather than on workmanship. Today, most amber is simply polished to display the gem's natural beauty and warm glow.

Legends and Lore

In Greek mythology, amber was formed when Phaethon, son of Helios, the sun god, was killed by lightning. Grief turned his sisters to poplar trees; their tears were drops of amber.

Occurrences

Ninety percent of the world's gem-quality amber is found along the southeastern shores of the Baltic Sea. Floating "sea" amber from these deposits is dispersed around the Baltic's shores. Most Baltic "pit" amber is mined from blue glauconite sand, called "blue earth." The largest deposits in this area are in the Samland Peninsula near Kaliningrad in Russia and around Gdańsk in Poland. The second most important source is the Dominican Republic (referred to as copal, and not an amber, by some). Other occurrences are in Sicily (simetite), Myanmar (Burma; burmite), Romania (romanite), Lebanon, Germany, Canada, and Mexico.

Evaluation

The best-quality amber is clear, transparent, and flawless. The most valuable colors are greens, blues, and reds. Of the common colors, yellow is the most highly prized. Pressed amber or amberoid is made by heating small pieces of amber and hydraulically compressing them into blocks. Amberoid is distinguished from true amber by its flow structure and the elongation of air bubbles. Imitation amber is made with plastics, modern natural resins, or glass.

OPPOSITE: Chinese carving of amber 10.9 cm (4¼ in) high, a string of 108 beads from the Baltic coast, and an irregular polished piece from Sicily, 11.5 cm (4½ in) long.

Coral

The orange-to-red gem often seen as Italian "horn" good-luck charms was long thought to be a sea plant with flowers but no leaves or roots. In 1723, the French biologist J. A. Peyssonnel identified coral as the exoskeleton of colonial polyps, small animals that create their dendritic forms from calcite dissolved in seawater. Although coral is a potentially renewable resource, reckless exploitation has placed the corals in jeopardy of extinction. Conservation efforts, initiated in the 1970s, aim at selective harvesting to preserve the gem corals; however climate change and its effect on the oceans has become a present threat.

CORAL DATA

Coral is formed primarily of either calcite, $CaCO_3$, or conchiolin, a horny organic substance.

Cleavage: None

Hardness: 3.5–4

Specific gravity: 2.6–2.7 (red calcite coral); 1–3 (black, gold, blue conchiolin coral)

R.I.: Not measurable

ABOVE: Nineteenth-century Chinese coral carving 35.5 cm (14 in) high.

Properties

Red color is the distinctive attribute of the traditional gem coral, although it ranges from red through orange to pink and even white. Red is due to iron and organic pigments, and fine skeletal structure renders coral opaque. The most valued gem coral is created by the coelenterate species *Corallium rubrum*, a hard coral. Black coral, known as "akbar" or "king's coral," golden coral, and the rare gray blue "akori" are soft corals. The lengthwise striped or patterned skeletal structure of their branches distinguishes all corals from imitations.

Historic Notes

Coral has been found with Paleolithic remains in Wildscheuer Cave, north of Wiesbaden, Germany. It was depicted on a Sumerian vase of about 3000 BCE. Coral was popular with the ancient Greeks and Romans. Pliny, writing in the first century CE, mentions coral trade between the Mediterranean countries and India. In the thirteenth century, Marco Polo noted coral in jewelry and adorning idols in Tibetan temples. Chinese mandarins wore coral buttons of office. The Spaniards introduced coral to Mesoamerica during the sixteenth century, and the Navajos and Pueblos used it extensively in jewelry.

The Victorians favored coral, and it was a favorite of Art Deco jewelers. Today, coral enjoys great popularity, but the supply will probably decrease or even cease if conservation efforts are unsuccessful.

Legends and Lore

Coral in Greek mythology originated with Medusa's death at the hands of Perseus; the drops of her blood became red coral. Coral amulets were thought to protect children from danger during Roman times. In *Metamorphoses*, Roman poet Ovid (ca. 43 BCE–17 CE) praised it as a cure for scorpion and serpent bites. Coral promotes good humor, according to Arab physician Avicenna (980–1037 CE). A twelfth-century English manuscript recommends coral engraved with a gorgon or serpent as protection against all enemies and wounds. A medieval English belief was that a coral necklace helped in childbirth. In Italy to this day, coral is worn as protection against the "evil eye."

Perseus Turning Phinneas to Stone with the Head of Medusa, drawing by Charles Monnet, 1767; in Greek mythology, the blood drops from Medusa's severed head hardened into coral in the Red Sea.

Canary Islands, off the northwest coast of Africa, and around Japan, southern Malaysia, Mauritius, Australia, and Taiwan.

Two species of black coral (*Antipatharia*) were discovered in 1957 off Maui; this coral is also found off Australia and the West Indies. And a pink coral, previously known from other regions, was discovered in 1966 off Oahu. Gold coral is the rarest of the Hawaiian corals and varies in color from gold, brownish gold, bamboo beige, or brown to dark olive green. Trade in many coral varieties is legally limited by the Convention on International Trade in Endangered Species of Wild Fauna and Flora (CITES).

Evaluation

Color, size, and polish determine coral's value. Pale rose ("angel skin") and deep red ("ox blood") are the most valued, and coral is sometimes stained to produce a more valuable shade. Large pieces are rare, and large, fine carvings command high prices. Necklaces are evaluated by matching color and evenness of the beads.

GEM CORAL IMITATIONS

Coral is imitated with conch pearl, conch shell, and powdered marble compacted under pressure. Plastics, wood, and sealing wax are also used. The Gilson coral, produced in orange to red colors, is an excellent recent imitation.

ABOVE LEFT: An illustration of coral skeletons from Henri de Lacaze-Duthiers, *Histoire naturelle du corail...*, 1864. American Museum of Natural History Research Library

Occurrences

Precious coral grows in clear, warm water at depths from 30 to 900 feet (10 to 300 m). The sources of the finest coral from early times have been the Mediterranean and Red Seas; the center of the coral industry, Torre del Greco, south of Naples, Italy, is also known as the City of Coral. The same or similar varieties grow in the Bay of Biscay, around the

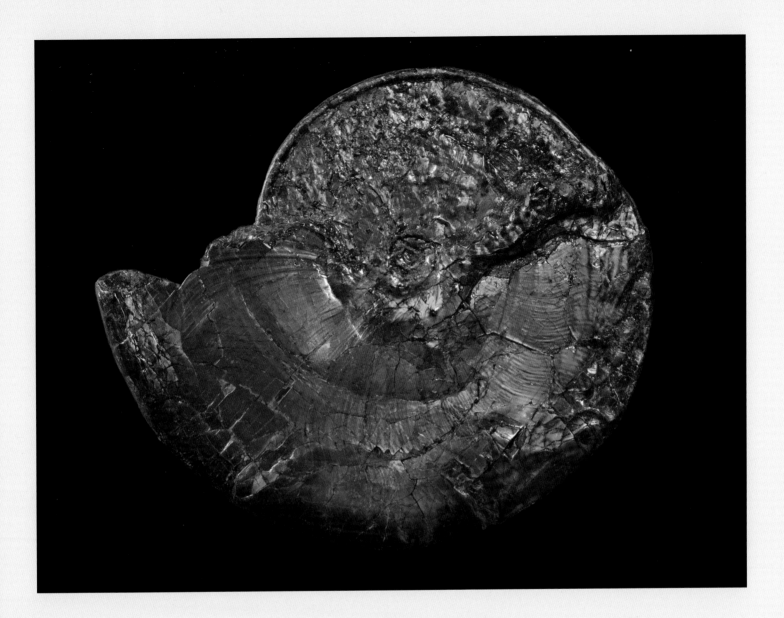

AMMONITE

Ammonite fossils are an incredible source of information for scientists, from dating rocks to confirming the presence of prehistoric seas. These extinct marine animals thrived in the Paleozoic and Mesozoic Eras, some 400 to 65 million years ago.

The vivid iridescent colors of this fossil ammonite is unique to those from Alberta, Canada, and adjacent Saskatchewan and Montana. The original "mother of pearl" lost the organic component (conchiolin) after burial, but the encapsulating sediment prevented the change of aragonite to calcite, so the platy structure and iridescence only increased upon fossilization.

This ammonite is approximately 56 cm (22 in) in diameter. These extinct marine animals thrived in the Paleozoic and Mesozoic eras, some 400 to 65 million years ago. Its vivid coloration is unique to ammonites from Alberta, Canada. Its iridescence is probably due to high temperatures and pressures during fossilization.

Jet

Though jet has an ancient tradition and was a charm of good fortune, not long ago it was associated with mourning jewelry. Queen Victoria wore it for forty years following Prince Albert's death in 1861, raising jet's popularity to its pinnacle. But by the early twentieth century, fashion changed, and the jet jewelry industry vanished. Today, with black in vogue, jet's popularity has revived. The term *jet-black* as a description of ultimate darkness is derived from this gem material.

♢ JET DATA

Composition: Carbon plus various hydrocarbon compounds

Cleavage: None, but brittle

Hardness: 3–4

Specific gravity: 1.3–1.35

R.I.: About 1.66

ABOVE: A polished jet slab 9.5 cm (3¾ in) long, a faceted jet stone of 2.92 cts., and an oval cabochon of 5.26 cts. All from unknown localities. The jet and turquoise frog from Chaco Canyon in New Mexico, measuring 8.1 cm (³⁄₁₆ in) in height, is from the Department of Anthropology at the Museum.

Properties

Jet is a dark brown to black variety of lignite (derived from the Latin *lignum*, meaning "wood"), a low-grade coal. This makes jet a rock rather than a mineral, or at best a mineraloid—a mineral-like substance. Jet will burn. It takes a high polish but scratches and abrades easily. It is sufficiently tough to be carved and faceted; a softer, brittle and less workable variety is called "bastard jet." Rubbed vigorously on wool or silk, jet develops an electric charge and attracts small pieces of straw or paper. This similarity to amber earned it the name "black amber."

Historic Notes

Usage of jet in Britain dates from at least the Neolithic; it was used as beads in the Bronze Age. Jet jewelry from Britain dates from the middle of the second millennium BCE, and the area around Whitby on the northeast coast of England has been the major source of the finest jet in the world. During Roman times, jet mining in Britannia was active, and a significant amount of jet jewelry made in Eboracum (York) was shipped to Rome. According to Pliny, the material was named for the town and river Gagas in Lycia (Turkey), where it or a similar substance was found.

Jet carving flourished in Spain during the fourteenth and fifteenth centuries, when jet was used in talismans and during periods of mourning. Pre-Columbian Mayas, Aztecs, Pueblos, and Native Alaskans used jet as decoration. The eighteenth and nineteenth centuries saw jet's extensive use in rosaries, crosses, carvings, and jewelry.

Legends and Lore

Jet has been considered protective, a gem for seafarers. It drives away venomous beasts, according to a book written in 1213 by Ibn al-Baitar, an Arabian botanist. Since the tenth century, Spanish jet *higas*, hand-shaped talismans, have been worn as protection from the evil eye.

Occurrences

Jet is found in lenticular masses embedded in hard black bituminous shale, known as "jet rock," where it formed by the lithification of submerged driftwood in seafloor mud. Jet is found in Germany, Spain, France, Poland, the United States, Russia, and India, as well as in England.

Evaluation

Uniform color and texture are the major factors to be considered. The compact homogeneous hard types take better polish and are considered the finest quality. Jet is moderately priced.

GEMSTONES CONFUSED WITH JET; JET SUBSTITUTES AND IMITATIONS

Obsidian, dyed chalcedony, and black tourmaline can be confused with jet.

Scotch cannel coal and Pennsylvania anthracite have been used as substitutes.

Imitation jet is made with glass, plastics, and vulcanite (hard vulcanized rubber). Black glass stones are often known as "Paris jet."

RARE & UNUSUAL GEMSTONES & ORNAMENTAL MATERIAL

Many minerals—for lack of sufficient abundance, uniformly good properties, or a popular tradition—do not rank among the better-known gems. Some are beautiful but not suited for use in jewelry and are mainly of interest to collectors. In this chapter, gem minerals have been segregated from the carving materials; this separates the facetable crystals from the rocks or "ornamental material." These gem minerals, often called "the rare and unusual gemstones," have been sorted into three categories: (1) minerals that have excellent properties but are too rare or are not so rare but have only adequate properties—usually lacking sufficient color or brilliance; (2) gem crystals that are too soft or fragile to be anything other than collectors' stones or part of a museum exhibition; and (3) opaque metallic minerals that have been faceted and fashioned or used in jewelry. The ornamental materials are listed last.

The entries in each group are ranked in descending order of "gem quality": a somewhat subjective evaluation. (See table, pages 188–89, for specific data on each mineral; some rare ones are also included only in this table.) We fully acknowledge our own biases in this ranking.

Very Rare and Moderately Good Gemstones

Z o i s i t e was first described in 1905, and the pink variety, thulite, particularly from Norway, was used as an ornamental stone for cabochons and carvings. In 1967, a new, intensely blue variety from Tanzania was named *tanzanite* by Henry B. Platt, vice president of Tiffany & Co., the firm that created a market for this gem. The crystals are transparent, sapphire blue to amethyst violet, and very strongly pleochroic. Some tanzanite is heat-treated to eliminate yellow or brown tinges and deepen the blue color. Because of its magnificent color and beauty, tanzanite has become popular as a faceted gem but fluctuates in availability and price with legal and mining issues in Tanzania.

B e n i t o i t e was discovered in San Benito County, California, in 1907 and established in 1985 as that state's gem, since it is found nowhere else. This rare gem has the color of blue sapphire and the dispersion of diamond.

LEFT: A 3.57-ct. benitoite gem and a crystal measuring 2 cm (3¾ in) across, with neptunite crystals in natrolite matrix. Both are from San Benito County, California.

OPPOSITE ABOVE: A kunzite crystal 9.8 cm (3¹³⁄₁₆ in) long and two cut stones of 121.48 and 191.84 cts. from Pala, California.

Spodumene varies in color from colorless to yellow, yellow green, pink, violet, pale to deep green, and pale green blue. Kunzite and hiddenite are the two most popular gem varieties. Kunzite is pink, lilac, or violet and was named after the famous gemologist George F. Kunz. Some kunzites fade on prolonged exposure to sunlight. Major sources are California, Brazil, Madagascar, and Afghanistan. Hiddenite is a rare gem restricted in occurrence almost exclusively to Hiddenite, North Carolina.

Kornerupine can be colorless, brown, and yellow, but green is the most valued color. Cat's eye and star kornerupines are very rare. Gem-quality material was first found in Madagascar in 1911; Sri Lanka, eastern Africa, Australia, Myanmar, Canada, and South Africa are also sources.

Sinhalite was identified in 1952 and named after Sinhala, the ancient Sanskrit name for Sri Lanka, where the gem was found. It also occurs in Myanmar and Tanzania

179

(pink to brownish pink). Previously, sinhalite was considered as brown peridot. Its typical colors are yellowish brown, greenish, and very dark brown.

Rutile is a common titanium-rich mineral. Its color is usually dark red or reddish brown to black. It is characterized by high refractive indexes and very strong dispersion. Its fire exceeds by six times that of diamond but is usually masked by dark colors. The rarity of transparent material and the darkness of the colors restricts its use to collectors.

Euclase is usually transparent and varies from colorless to assorted colors; blue is the most prized. The name derives from the Greek *eu*, meaning "easy," and *klasis*, meaning "fracture," in allusion to the mineral's perfect cleavage. If the supply were not so limited, it could be a popular gem; Minas Gerais, Brazil, is a major source for gem-quality euclase. Other sources include Myanmar, Colombia, Tanzania, and Zimbabwe.

Titanite, also known as sphene, is yellow, brown, or green and in gem quality is transparent. It has high refractive indexes and strong dispersion, giving the well-cut stones high brilliance and fire. It would be an important gem mineral if it were harder and less brittle. Gem titanite is found in Madagascar, Myanmar, India, Kenya, Tanzania, Sri Lanka, Brazil, and Mexico.

Diopside is a member of the pyroxene group of rock-forming minerals, seldom found in gem quality. Occasionally, it is cut into faceted stones, cat's eyes, and four-rayed star stones. It most commonly occurs in different shades of green. Myanmar, Russia, Italy, and New York are important sources of gem diopside.

Obsidian is the most important of the natural glasses for use in jewelry. Obsidian is a transparent to opaque volcanic glass; it is usually black but may also be brown, green, yellow, red, or blue. Occasionally, it exhibits a golden or silver iridescent sheen caused by reflection from tiny inclusions.

OPPOSITE: A group of color-zoned euclase crystals, the largest of which is 5.5 cm (2⅛ in) long, from Zimbabwe; and two cut gems of 7.94 cts. and 8.64 cts., from Minas Gerais, Brazil.

RIGHT: Titanite: a 10.07-ct. stone from Switzerland and a twinned crystal from Austria, 5.5 cm (2⅛ in) long.

Scapolite is actually a mineral group and may be colorless, pink, violet, yellow, or gray. It makes attractive cat's eyes and faceted stones. Gem-quality scapolite was first found in 1913 in Mogok, Burma (Myanmar), but the Madagascar and Tanzania are also sources.

Amblygonite of gem quality is usually yellow, greenish yellow, or lilac. The relative rarity and pale colors

Snowflake, or flowering, obsidian is a black variety with white inclusions. Mahogany obsidian is a banded black and red variety. Apache tears are small rounded pebble-like pieces, usually translucent and light to dark gray in color, found in the American West. Major occurrences are worldwide.

of amblygonite restrict its use in jewelry. The major sources of gem amblygonite are Brazil, Myanmar, and Maine in the United States.

ABOVE: Illustration by Pietro Fabris of volcanic rock samples and jeweled pin, reproduced in Sir William Hamilton, *Campi Phlegraei: Observations on the Volcanoes of the Two Sicilies* (Naples: [s.n.] 1776), from the American Museum of Natural History Research Library.

RIGHT: Amblygonite cut stone of 34.00 cts. and an irregular crystal of gem-quality amblygonite 8 cm (3⅛ in) high, both from Minas Gerais, Brazil.

Soft and Fragile Gemstones

Calcite is the most common carbonate mineral and is very abundant. Aragonite is chemically identical to calcite but has a different crystal structure. Faceting calcite is difficult because it has perfect cleavage in three directions. Calcite may be colorless, white, gray, red, pink, green, yellow, brown, or blue. Iceland spar is the transparent colorless variety of calcite. Gem-quality calcite is found in many localities, particularly Mexico. Marble, the popular stone used for statues and decorative objects, is a metamorphic rock consisting predominantly of calcite. Onyx marble is banded calcite and/or aragonite. Its major source is Baja California in Mexico, and hence it is occasionally called "Mexican onyx" or, if dyed green, "Mexican jade." There are sources in Utah.

Fluorite is fragile but occurs in a wide variety of colors. *Fluorescence* derives its name from this mineral, which displays the luminous property vividly. Because of its attractive colors, it is occasionally faceted for collectors. The variety banded in white and blue, violet, or purple—known as Blue John or Derbyshire spar—has been used since Roman times for ornamental objects. Southern Illinois is a major source of fluorite.

ABOVE: Calcite crystal 7.3 cm (2⅞ in) long and a 99.6-ct. gem from Gallatin County, Montana.

Rhodochrosite has been used commercially for decorative objects, beads, and cabochons since a beautifully banded massive material was discovered at San Luis in Argentina before World War II. Rhodochrosite from Argentina is occasionally referred to as "Inca Rose" because the Incas worked the same deposits. Rhodochrosite is also found as deep red crystals; these are occasionally faceted for collectors—particularly crystals from Hotazel, South Africa. China is a relatively new source.

OPPOSITE: A 59.65-ct. rhodochrosite gem from Kuruman, South Africa, the largest on record, and a group of crystals that is 7 cm (2¾ in) wide.

ABOVE: A pyrite specimen from Chester, Vermont, approximately six inches in diameter.

Metallic Opaque Gemstones

Pyrite, known as "fool's gold," is the most common sulfide mineral and occurs throughout the world. When used in jewelry, it has been called "marcasite." This is a misnomer; marcasite, though chemically identical, is a mineral with a different crystal structure. Pyrite was used by the ancient Greeks and Romans, Mayas, Aztecs, and Incas. Its popularity revived in the late 1980s. Pyrite is opaque with a brass-yellow color and bright metallic luster.

Hematite, one of the most important ores of iron, is black to dark gray with a metallic luster. It is fashioned into intaglios, cameos, and occasionally beads imitating black pearls or faceted stones often sold as black diamonds. There are many sources; Alaska is now a major one.

Ornamental Material—Carvings, Beads, Inlays

Gypsum has three varieties that have been used as ornamental stones since ancient times. *Alabaster* is the massive, fine-grained, translucent variety. *Satin spar* is the fibrous variety with a pearly luster. Selenite is the transparent colorless crystal form. *Gypsum* is very soft and can be scratched with a fingernail. It is usually white, but it may also be yellowish, brownish, reddish, or greenish. The massive variety is porous and is easily dyed. The most important sources of alabaster are Tuscany, Italy, and Derbyshire and Staffordshire, England.

Talc, when free from admixture, is silvery white but with impurities becomes gray, green, reddish, brown, or yellow. It is the softest gem mineral. *Steatite*, a popular material for carvings, is massive talc containing impurities that increase its hardness. It has a greasy, soapy feel and is also known as soapstone. The translucent material has a higher value than the opaque. *Agalmatolite* is a brownish variety of steatite. Steatite occurs at many locations.

Pyrophyllite is a rare mineral occasionally used for cabochons and more frequently for carvings. It is soft and usually opaque, has a pearly to greasy luster, and comes in colors varying from white to gray, pale blue, and brown. Translucent material is the most valued. A considerable part of the so-called agalmatolite, commonly used in Chinese carvings, is compact pyrophyllite. An important source is China.

Malachite is a vivid green copper mineral widely used for cabochons, beads, carvings, and inlaid work. It was known in Egypt as early as 3000 BCE and was used for amulets, jewelry, and as powder for eye shadow. In the early nineteenth century, the famous Ural mines were highly productive and supplied malachite to Europe. It was worn in Italy as an amulet against the evil eye. Malachite seldom occurs in visible crystals, but is usually massive or fibrous. The massive variety, banded in green hues, is the most attractive for jewelry. Malachite is soft, brittle, and sensitive to heat, acids, and ammonia—not sufficiently durable for ring stones. Currently, the major producer of malachite is the Democratic Republic of the Congo.

Azurite is a soft, opaque gem of an intense azure blue. Most often cabochons, beads, or decorative objects are fashioned from massive azurite. Important sources are Arizona, Mexico, and China.

Rhodonite in its massive form is popular for cabochons, beads, vases, boxes, goblets, and other decorative objects. It was first used in Russia in the late eighteenth century. Rhodonite has an attractive rose red color, usually with black veins of manganese oxides. Major sources are Japan and the Ural Mountains of Russia.

ABOVE: A 4,000-kg (4½-ton) block of azurite-malachite 1.5 m (5 ft) tall from the Copper Queen Mine in Bisbee, Arizona.
OPPOSITE: Chinese malachite vase, 20 cm (7⅞ in) high.

Rare and Unusual Gems

SPECIES Variety	CHEMICAL FORMULA	HARDNESS	SPECIFIC GRAVITY	REFRACTIVE INDEX	COMMENT
VERY RARE GEMSTONES					
Zoisite Tanzanite: a beautiful blue pleochroic variety, rare Thulite: a massive pink ornamental stone	$Ca_2Al_3Si3O_{12}(OH)$	6–6.5	3.15–3.38	1.685–1.725	
Benitoite	$BaTiSi_3O_9$	6–6.5	3.64–3.7	1.757–1.804	Sapphire blue, strong fire; very rare
Spodumene Kunzite: lilac colored, some popularity Hiddenite: emerald green, exceedingly rare	$LiAlSi_2O_6$	6.5–7	3.0–3.2	1.66–1.676	Good cleavage in two directions
Kornerupine	$Mg_3Al_6(Si,Al,B)_5O_{21}(OH,F)_2$	6–7	3.28–3.35	1.661–1.669	Green to brown; rare
Rutile	TiO_2	6–6.5	4.2–4.3	2.62–2.9	Dark and somewhat soft
Sinhalite	$MgAlBO_4$	6.5–7	3.47–3.5	1.665–1.712	Mistaken for brown peridot
Euclase	$BeAlSiO_4(OH)$	6.5–7.5	3.05–3.1	1.65–1.676	One perfect cleavage; rare
Titanite	$CaTiSiO_5$	5–5.5	3.44–3.55	1.843–2.11	Fine brilliance and fire but soft
Diopside	$CaMgSi_2O_6$	5–6	3.2–3.3	1.664–1.721	Green pyroxene; rare in multi-carat sizes
Obsidian	Natural glass	5–5.5	2.4	1.48–1.51	Usually dark, brittle
Scapolite	$(Ca,Na)_4Al_3(Al,Si)_3Si_6O_{24}(Cl,CO_3,SO_4)$	5–6	2.5–2.74	1.539–1.579	In various colors and cat's eyes
Amblygonite	$(Li,Na)AlPO_4(F,OH)$	5.5–6	3–3.1	1.578–1.619	Various pale colors; rare
SOME COLLECTOR GEMSTONES					
Calcite Iceland spar— optically flawless colorless calcite	$CaCO_3$	3	2.7	1.486–1.658	High birefringence, soft, perfect cleavage in three directions
Fluorite	CaF_2	4	3.18	1.434	Many colors, soft, octahedral cleavage
Rhodochrosite	$MnCO_3$	3.5–4.5	3.45–3.6	1.597–1.1817	Soft with three cleavages— massive form ornamental stone
Barite	$BaSO_4$	3–3.5	4.3–4.6	1.636–1.648	Various colors but soft and one perfect cleavage

SPECIES Variety	CHEMICAL FORMULA	HARDNESS	SPECIFIC GRAVITY	REFRACTIVE INDEX	COMMENT
METALLIC GEMSTONES					
Pyrite	FeS_2	6–6.5	5.02		Brassy, called marcasite in marketplace
Hematite	Fe2O3	5.5–6.5	5.26		Steely black, called black diamonds
ORNAMENTAL MATERIALS					
Gypsum Alabaster—massive fine-grained rock Satin spar—fibrous with pearly luster Selenite—transparent colorless crystal	$CaSO_4 \cdot 2(H_2O)$	2	2.3	1.52–1.53	Soft and abundant
Calcite Onyx—massive fine-grained rock	$CaCO_3$	3	2.7	1.486–1.658	Various colors
Talc Steatite or soapstone—massive fine-grained rock, often apple green	$Mg_3Si_4O_{10}(OH)_2$	1	2.2–2.8	1.54	Soft and greasy feeling
Pyrophyllite Agalmatoline—creamy white to brown massive form	$Al_2Si_4O_{10}(OH)_2$	1–2	2.65–2.9	1.58	Soft, resembles talc
Malachite	$Cu_2CO_3(OH)_2$	3.5–4.5	3.6–4.1	1.85	Light to dark green, banded and massive
Azurite	$Cu_3(CO_3)_2(OH)_2$	3.5–4	3.77	1.73–1.836	Dark azure blue, alters to malachite
Rhodonite	$MnSiO_3$	5.5–6.5	3.57–3.76	1.73	Rose, pink to brownish red, usually massive
Variscite	$AlPO_4 \cdot 2H_2O$	3.5–4.5	2.2–2.57	1.56	Massive blue green mistaken for turquoise
Serpentine Bowenite—green jade-like rock	$Mg_3Si_2O_5(OH)_4$	4–6	2.44–2.62	1.56	Commonly in soft green rock serpentinite

Acknowledgments

There are many people to whom the authors are greatly indebted for their efforts, assistance, and advice, both for the previous edition and the new revised edition.

We doubt you would be reading this book at all without the dedication, ministrations, and good humor of Nancy Creshkoff, our original stylist and editor.

The moral support, valuable advice, and infinite patience of Tom Sofianides and Carole Slade, our spouses, were imperative to our completing this book.

Without the participation of Joseph J. Peters, including his painstaking review of information on the collections and assistance with many aspects of this project, it would not have come to fruition.

We appreciate the patience and assistance of others in the Museum's Department of Mineral Sciences, particularly Janice Yaklin and Charles Pearson, and at *Natural History*, L. Thomas Kelly and Scarlett Lovell. Thank you also to Jill Hamilton, for proofreading.

The Museum's photographic studio was generous with its assistance, and we thank Jackie Beckett, Kerry Perkins, and Denis Finnin.

Many people have been generous with information: Joan Aruz, Wendy Ernst, Linda Eustis, Leonard Gorelick, Eugene Libre, George Morgan, James Pomarico, Frank Rieger, Peter Schneirla, David Seaman, James Shigley, and Nicholas Steiner.

For the previous edition, helpful people include Gaston Giuiliani, Richard Hughes, Robert Kane, Neil Landman, and William Larson. For the new edition, we would like to thank the team at Sterling Publishing, including Barbara Berger, executive editor; Chris Thompson, art director, interiors; Elizabeth Lindy; art director, covers; David Ter-Avanesyan, cover designer; Fred Pagan, production manager; and Marilyn Kretzer, editorial director. Also special thanks to Lary Rosenblatt at 22MediaWorks, and Fabia Wargin Design.

We are grateful to those who have helped create the mineral and gem collections. Those whose donations are featured in this book are as follows:

George Ackerman; Mrs. R. T. Armstrong; Mrs. Frank L. Babbott; Lilias A. Betts; Mrs. C. L. Bernheimer; Susan D. Bliss; Maurice Blumenthal; David A. Byers; Elizabeth Varian Cockcroft; Lawrence H. Conklin; Joseph F. Decosimo; Dr. B. Delavan; Mrs. George Bowen DeLong; Lincoln Elsworth; Alexander J. and Edith Fuller; Leila B. Grauert; Jack Greenberg; Peter Greenfield; K. B. Hamlin; Dr. George E. Harlow; Mrs. William H. Haupt; Lloyd Herman; Dr. Maurice B. Hexter; Mrs. Charles C. Kalbfleisch; Morton Kleinman; Dr. George E. Kunz; Korite Minerals, Ltd.; Vincent Kosuga; Mabel Lamb; Charles Lanier; Mrs. Zoe B. Larimer; Gerald Leach; S. Howard Leblang; Mrs. Bonnie LeClear estate; Vera Lounsbery estate; Roy Mallady; Alastair Bradley Martin; Mrs. Patrick McGinnis; Roswell Miller Jr.; Milton E. Mohr; Dr. Arthur Montgomery; John Pierpont Morgan Sr.; John Pierpont Morgan Jr.; M. L. Morgenthau; Dr. Walter Mosmann; Dr. Henry Fairfield Osborn; Clara Peck estate; Phelps-Dodge Corporation; Dwight E. Potter; Arthur Rasch; Dr. Julian Reasonberg; Robinson & Sverdlik Company; Mr. and Mrs. J. Robert Rubin; Hyman Saul; Elizabeth Cockcroft Schettler; Mr. and Mrs. Bernard Schiro; Dr. Louis Schwartz; Victoria Stone estate; William Boyce Thompson estate; John Van Itallie; David Warburton; Thomas Whiteley; and several anonymous donors.

Glossary

Allochromatic Pertaining to color resultant from a mineral impurity, such as minor chemical substitutions or radiation damage.

Alluvial deposit A density-driven concentration by water moving in a river or stream.

Amulet *See* Talisman.

Asterism Chatoyancy in two or more directions, giving a starlike appearance in illumination.

Axis (or crystal axis) A reference direction in a crystal that is parallel to symmetry directions or the intersection of faces.

Birefringence The magnitude of the difference in the refractive indices (R.I.s) of birefringent minerals.

Birefringent (or doubly refractive) Having two or three R.I.s, a characteristic of minerals not possessing cubic symmetry.

Brilliance Degree to which a faceted gem sparkles and returns light from within; dependent upon cut and R.I. Synonyms: life, liveliness.

Cabochon A gem-cut style distinguished by its smooth convex top and no facets.

Carat The standard unit of gem weight (mass); 1 carat = 0.2 grams.

Chatoyancy Cat's eye appearance when a rounded stone is illuminated. Caused by parallel arrangement of tiny needles within a crystal or a coherent mass of parallel fibers (as in cat's eye nephrite).

Cleavage The tendency of a mineral to break along a plane due to a direction of weakness in the crystal.

Cryptocrystalline Constituted of submicroscopic crystals.

Crystal A solid body having a regularly repeating arrangement of its atomic constituents; the external expression may be bounded by natural planar surfaces called "faces."

Crystalline Having the properties of a crystal: a regular internal arrangement in three dimensions of constituent atoms.

Cubic (crystal system) Defined by three mutually perpendicular axes of equal length— the highest symmetry class.

Dichroism Pleochroism in two directions.

Dispersion The systematic variation of refractive index with color in a substance; colors separate during refraction of white light. It leads to fire in a gem. It is expressed numerically as the difference in refractive index between the G and B Fraunhofer lines in sunlight at 430.8 and 686.7 nanometers.

Fire Division of colors in a colorless transparent gem such as diamond; due to dispersion.

Fraunhofer lines Light from the Sun includes distinct sharp dark wavelength bands due to characteristic absorptions by chemical elements like oxygen, hydrogen and sodium in the cold gases around the Sun. Named for physicist Joseph von Fraunhofer.

Gem A mineral (gemstone) that has been fashioned to enhance its natural beauty.

Gemstone A substance that has beauty, durability, and rarity and that can be fashioned into personal adornment.

Group (mineral group) A set of minerals that share the same crystal structure.

Habit A characteristic shape of a mineral, either a crystal shape or the shape and style of polycrystal-line intergrowths.

Hardness Resistance to scratching; measured from 1 to 10 on the Mohs scale.

Hexagonal (crystal system) Defined by three equal axes lying in a plane and intersecting at 120-degree angles and a fourth perpendicular axis that is a sixfold rotation.

Idiochromatic Color is inherent and due to some aspect of chemical composition and crystal structure.

Imitation A substance that simulates a genuine gem, typically glass, plastic, and other noncrystalline materials.

Intergrowth A composite of crystals in intimate contact.

Iridescence Color produced by light interference, as in labradorite feldspar.

Luster The manner in which a substance reflects light from its surface; affected by the surface's smoothness, the substance's reflectivity, and any layering structures, such as in a pearl.

Magma Mobile molten rock material from which igneous rocks form by solidification.

Mineral A naturally occurring substance (usually inorganic) that is crystalline and has a composition that can be defined by a simple chemical formula.

Monoclinic (crystal system) Defined by three nonparallel axes where there are only two right angles between the axes and no high-order rotation axes.

Orthorhombic (crystal system) Defined by three unequal mutually perpendicular axes.

Pegmatite (gem) An igneous rock with conspicuously large mineral grains and often enriched with volatile elements in minerals such as beryl (Be), spodumene (Li), topaz (F), and tourmaline (B).

Piezoelectric Capable of producing a surface electric charge when deformed elastically; a property of some minerals without a center of symmetry.

Placer deposit An accumulation of dense mineral grains at the bottom of a sediment pile by the weathering action of a moving fluid such as water in a river (alluvial deposit) or along a beach or wind.

Play of colors A range of colors seen in a gemstone such as opal when it is viewed from different angles; due to optical diffraction.

Pleochroism The phenomenon whereby the color intensity or the actual color of a crystal is different depending on the orientation in which it is observed.

Pseudochromatic Coloring due to physical causes such as dispersion or foreign included particles and internal boundaries.

Pyroelectric Capable of producing a surface electric charge when temperature changes; a property of some minerals that do not have a center of symmetry.

Refraction The bending of light (or any wave phenomenon) when it moves between media with different conductive velocities.

Refractive index (R.I.) A mathematical constant equal to the ratio of the velocity of light in a substance to that in a vacuum; it determines the angle at which light bends when it enters a substance obliquely. There can be as many as three values in minerals of lower symmetry, and the observed value varies with with illumination direction to the crystal axes. The numerical value is usually measured in the yellow light (589.3 nanometer wavelength) of a sodium-vapor lamp.

Rock A consolidated assemblage of grains of one or more minerals.

Rough The raw gemstone.

Schist A metamorphic rock having a subparallel alignment of the principal constituent mica or micalike (platy) minerals.

Silica Silicon dioxide; used to inform about chemical composition without reference to a solid substance, such as quartz.

Simulant A substance used to simulate a gemstone, usually a synthetic material with a similar appearance to the simulated gemstone.

Sixling A twin intergrowth of six crystals that appears to have hexagonal symmetry; a common habit for chrysoberyl.

Specific gravity A dimensionless measure of density (numerically equivalent to the value in grams per cubic centimeter–density).

Symmetry The correspondence in shape or length of elements of a body; as repeated by a mirror, rotation about an axis, or inversion through a point (center of symmetry).

Synthetic An artificial substance that is identical to a natural one.

Talisman An object, sometimes fashioned and engraved with a symbol, that is believed to provide magical, medicinal, or protective power. Synonym: amulet.

Tetragonal (crystal system) Defined by three mutually perpendicular axes, two of which are of equal length.

Trigonal (crystal system) Defined by three equal axes lying in a plane and intersecting at 120-degree angles and a fourth perpendicular axis that is a threefold rotation axis.

Trilling A twin intergrowth of three crystals that appears to have trigonal symmetry.

Twin (twinned crystal) A nonparallel intergrowth of separate crystals related by symmetry not possessed by the substance.

Variety (gemstone) A named specific color or other quality of a gemstone species, such as ruby for red corundum.

Volatiles (components) In magma, those materials that readily form a gas and are the last to enter into and crystallize as minerals during solidification.

OPPOSITE: Naturally colored diamonds from the Aurora collection (see page 25).

Reading List

Books

Arem, Joel E. *Color Encyclopedia of Gemstones*, 2nd ed. New York: Van Nostrand Reinhold, 1987.
Best compendium of gem data and account of largest stones in museums at the time of publication; many good photos.

Balfour, Ian. *Famous Diamonds*, 5th ed. Woodbridge, UK: Antique Collectors Club, 2009.
The definitive book on large famous diamonds.

Ball, Sydney H. *Roman Book on Precious Stones (Including an English Modernization of the 37th Booke of the Historie of the World by C. Plinius Secundus)*. Los Angeles: Gemological Institute of America, 1950.
If you want to know what Pliny had to say, this is the place to get it.

Bancroft, Peter. *Gem and Crystal Treasures*. Tucson, AZ: Western Enterprises/Mineralogical Record, 1984.
A good compilation of information on mineral and gem localities.

Bauer, Max. *Precious Stones*. New York: Charles E. Tuttle Company, 1982.
One of the references on gems found in a good library.

Bruton, Eric. *Diamonds*, 2nd ed. Radnor, PA: Chilton, 1978, 1993.
Excellent, authoritative, readable coverage of every aspect of diamonds.

Dietrich, R. V. *The Tourmaline Group*. New York: Van Nostrand Reinhold, 1985.
Still an authoritative book on tourmalines.

Evans, Joan. *Magical Jewels of the Middle Ages and the Renaissance*. Oxford, UK: Clarendon Press, 1922. (New York: Dover Publications, 1977).
A scholarly work that cites directly from original texts on gems from ancient times through the seventeenth century. The most important book of its kind.

——. *A History of Jewellery: 1100–1870*. Boston: Boston Book and Art, 1970.
An important, authoritative work.

Groat, L. A. (ed.). *The Geology of Gem Deposits*, 2nd ed. Short Course Handbook Series 44. Quebec: Mineralogical Association of Canada, 2014.
A compendium of chapters on the geology of the important gemstones.

Harlow, George. E. (ed.). *The Nature of Diamonds*. Cambridge, UK: Cambridge University Press, 1998.
An informative and authoritative work.

Heiniger, Ernst A., and Jean Heiniger. *The Great Book of Jewels*. New York: New York Graphic Society, 1974.
An opulent coffee-table book.

Hughes, R. W. *Ruby and Sapphire: A Collector's Guide.*
Bangkok: Gem and Jewelry Institute of
Thailand, 2014.
**The latest and greatest book by the authority
on corundum gems.**

Keller, Peter C. *Gemstones and Their Origins.* New York:
Van Nostrand Reinhold, 1990.
**One of the few books on the geological conditions of
gemstone formation.**

Klein, Cornelius, and Barbara Dutrow. *Manual of Mineral
Science*, 23rd ed. (Manual of Mineralogy). Hoboken,
NJ: John Wiley & Sons.
A standard textbook on mineralogy.

Kunz, George F. *The Curious Lore of Precious Stones.* Garden
City, NY: Halcyon House, 1938 (1913). (New York:
Dover Publications, 1971).
**A mixture of historical information and recounting
from numerous lapidaries; available as a free,
downloadable e-book at https://archive.org/details/
curiousloreofpre028009mbp.**

———. *The Magic of Jewels and Charms.* Philadelphia:
J. B. Lippincott, 1915.
More wonderful information from the great gem expert.

Landman, N. H., P. M. Mikkelson, R. Bieler, and B. Bronson.
Pearls: A Natural History. New York: Harry N.
Abrams, 2001.
A book with the allure and luster of its subject.

Muller, Helen. *Jet.* Oxford, UK: Butterworth-Heinemann,
1987.
A treasure of information on jet.

Ogden, Jack. *Jewelry of the Ancient World.* New York:
Rizzoli, 1982.
An important and definitive book on the topic.

Post, J. E. *The National Gem Collection.* New York:
Harry N. Abrams, 1997.
**A look at the great gem collection of the Smithsonian
Institution.**

Schumann, Walter. *Gemstones of the World*, 5th ed.
New York: Sterling Publishing, 2013.
**This is the latest edition of a standard reference
on all gemstones.**

Sinkankas, John, *Gemstones of North America*, vol. 2.
New York: Van Nostrand Reinhold, 1959.
**A bit dated but an excellent book; any book by John
Sinkankas is good.**

Webster, Robert, and E. Alan Jobbins. *Gemmologists'
Compendium*, 7th ed. London: N.A.G. Press, 1999.
A gemologist's handbook for the serious student.

Journals

Gems and Gemology
The quarterly of the Gemological Institute of America,
Carlsbad, CA.
The American journal on gemology.

In Color
The journal of the International Colored Gemstone
Association, Hong Kong.
An English language journal focused on colored gems, featuring articles on the market, gem sources, etc.

Journal of Gemmology
The journal of Gem-A, the Gemmological Association
of Great Britain, London.
The British journal on gemology.

Lapidary Journal
Chilton Publishing, Radnor, PA.
The amateur's journal, full of news, locality information, etc.

Mineralogical Record
Mineralogical Record Publishing, Tucson, AZ.
An amateur/professional journal on mineralogy with fine articles on localities, species, and reviews for mineral collectors.

Rocks and Minerals
Taylor & Francis Group, Philadelphia, PA.
An amateur/professional journal on minerals, rocks, and gems with well-written, illustrated articles for collectors and aficionados.

Online Resources

Gemdat.org
The gemstone and gemology information website.
http://www.gemdat.org/

Gemological Institute of America
Gems and Gemology journal and other information.
http://www.gia.edu/

International Colored Gemstone Association
Gem by gem information.
http://gemstone.org/

Citations

Abel, Eugenius. *Orphei Lithica Accedit Damigeron de lapidibus Recensuit Eugenius Abel.* Paris: Berolini, 1881.

Ball, Sydney H. "Historical Notes on Gem Mining." *Economic Geology* 26 (1931): 681–738.
An early and important effort on the history of mining.

Beasley, W. L. "The Morgan Gem Collection in the American Museum of Natural History." *Jewelers' Circular Weekly* (February 2, 1916): 88–95.

Boyle, Robert. "Some Considerations Touching the Usefulness of Experimental Natural Philosophy." (1663, 1671). Republished, London: Forgotten Books, 2013.
A source of early information on gems.

Evans, Joan. *Magical Jewels of the Middle Ages and the Renaissance*. Oxford, UK: Clarendon Press, 1922.

Federman, David. *Modern Jeweler's Gem Profile: The First 60*. Shawnee Mission, KS: Vance Publishing, 1988. **Information, stories, and news on sixty gemstones from the trade's perspective.**

Fritsch, E., and G. R. Rossman. "An Update on Color in Gems, Part 1: Introduction and Colors Caused by Dispersed Metal Ions." *Gems and Gemology* 24 (Fall 1987): 126–39.

——. "An Update on Color in Gems, Part 2: Color Involving Multiple Atoms and Color Centers." *Gems and Gemology* 25 (Spring 1988): 3–15.

——. "An Update on Color in Gems, Part 3: Colors Caused by Band Gaps and Physical Phenomena." *Gems and Gemology* 25 (Summer 1988): 81–102.

Giuliani, G., M. Chaussidon, C. France-Lanord, H. Savay-Guerraz, P. J. Chiappero, H. J. Schubnel, E. Gavrilenko, and D. Schwarz. "L'exploitation des mines d'émeraude d'Autriche et de la Haute Egypte à l'Epoque Gallo-romaine: mythe ou réalité?" *Revue de Gemmologie* 143 (2001): 20–24.

Giuliani, G., M. Chaussidon, H. J. Schubnel, D. H. Piat, C. Rollion-Bard, C. France-Lanord, D. Giard, D. de Narvaez, and B. Rondeau. "Oxygen Isotopes and Emerald Trade Routes Since Antiquity." *Science* 287 (2000): 631–33.

Giuliani, G., D. Ohnenstetter, A. E., Fallick, L. A. Groat, and A. J. Fagan. "The Geology and Genesis of Gem Corundum Deposits." In L. A. Groat (ed.), *Geology of Gem Deposits*. (Mineralogical Association of Canada Short Course), vol. 44, chapter 2 (2014): 29–112.

Gratacap, L. P. "The Collection of Minerals." *American Museum of Natural History Guide*, Leaflet 4 (1902): 1–21.

——. *A Popular Guide to Minerals*. New York: Van Nostrand Reinhold, 1912.

Groat, L. A., G. Giuliani, D. D. Marshall, and D. Turner. "Emerald Deposits and Occurrences: A Review." *Ore Geology Reviews* 34 (2008): 87–112.

Kunz, George F. "The Morgan Collection of Precious Stones." *American Museum Journal* 13, no. 4 (1913): 159–68, 171.

——. and Charles H. Stevenson. *The Book of the Pearl*. New York: Century, 1908. **Still an important book on pearls.**

Legrand, Jacques. *Diamonds: Myth, Magic and Reality*. New York: Crown, 1980. **An authoritative coffee-table book on diamonds.**

Liddicoat, Richard T., Jr. *Handbook of Gem Identification*. Santa Monica, CA: Gemological Institute of America, 1987. **The American source on gem identification.**

Meen, V. B., and A. D. Tushingham. *Crown Jewels of Iran*. Toronto: University of Toronto Press, 1968. **The incredible Iranian collection with excellent photos.**

Nassau, Kurt. *Gems Made by Man*. Radnor, PA: Chilton, 1980. **Best book by far on gemstone synthetics and substitutes. Includes good section on color.**

———. *Gemstone Enhancement.* Oxford, UK: Butterworth-Heinemann, 1980.
The authority on treatment but dated.

Pardieu, Vincent. "Hunting for 'Jedi' Spinels in Mogok." *Gems and Gemology* 50 (2014): 46–57. See: http://www.gia.edu/gems-gemology/spring-2014-pardieu-jedi-spinels-in-mogok.

———. "Amphibole Related Rubies from Mozambique: A Revolution in the Ruby Trade" (abs). In *21st General Meeting International Mineral Association (IMA) Abstr.* D. Chetty, L. Andrews, J. de Villiers, R. Dixon, P. Nex, et al. (eds.1. Johannesburg: Geological Society South Africa/Mineral Association South Africa, 2014: 281.

Pezzotta, Federico, and Brendan M. Laurs. "Tourmaline: The Kaleidoscope Gemstone." *Elements* 7, no. 5 (2011): 333–38.

Pliny the Elder. *Natural History.* vol. 10, books 36–37. D. E. Eicholz (trans.), T. E. Page et al. (eds.). Loeb Classical Library. London: William Heinemann, 1962. (See also Ball, *Roman Book on Precious Stones*, 1950.)

Pough, F. H. "Gem Collection of the American Museum of Natural History." *Gems and Gemology* 7, no (1) (1953): 323–34, 351.

Schubnel, Henry-Jean. *Color Treasury of Gems and Jewels.* London: Crescent Books, Orbis, 1971.
A book full of fine photos by a European expert.

Shigley, J. E., B. M. Laurs, A. J. A. Janse, S. Elen, and D. M. Dirlam. "Gem Localities of the 2000s." *Gems and Gemology* 46, no. 2 (2010): 188–216.

Sinkansas, John. *Emerald and Other Beryls.* Radnor, PA: Chilton, 1981.
The book on the subject.

———. *Gemology: An Annotated Bibliography.* 2 vols. Metuchen, NJ: Scarecrow Press, 1993.
Exactly what it says and comprehensive.

Webster, Robert. *Gems: Their Sources, Description, and Identification.* Hamden, CT: Archon Books, 1975.
Most comprehensive book on gems, although now dated.

Whitlock, Herbert P. *The Story of the Minerals.* New York: American Museum Press, 1925.

———. *The Story of the Gems.* New York: Lee Furman, 1936. (New York: Garden City Publishing, 1940).
The previous book on the Museum's gem collection and gems in general, although now dated.

Wodiska, Julius. *A Book of Precious Stones.* New York: G. P. Putnam's & Sons, 1906.
Another early compendium with data and history.

Woodward, Christine, and Roger Harding. *Gemstones.* New York: Sterling Publishing, 1988.
A good introductory book on gems, British flavor.

Yager, T. R., W. D. Menzie, and D. W. Olson. "Weight of Production of Emeralds, Rubies, Sapphires, and Tanzanite from 1995 through 2005" (2008., U.S. Geological Survey Open-File Report 2008–1013, 9 pp., available only online, http://pubs.usgs.gov/of/2008/1013.

Index

Note: Numbers in italics indicate photos and illustrations.

Picture Credits

Courtesy American Museum of Natural History Photographic Collection
viii, x, xii, xiii, xiv, xv, 7, 123, 125, 166, 173, 185, 186

Courtesy American Museum of Natural History Research Library
xvi–1, 5r, 172, 182t

Courtesy American Museum of Natural History
9, 19, 20, 21, 22–23

Courtesy Google Books
Anselmii Boëtii de Boodt ... Gemmarium et Lapidum Historia ... (Anselmus de Boodt), 1609: 5l; *Travels in India*, 1677 (Jean-Baptiste Tavernier): 71; *The Great Empress Dowager of China*, 1911 (Philip Walsingham Sergeant): 162

iStockPhoto
© boggy22: 92

Courtesy Library of Congress
Detroit Publishing Co.: ii–iii; Bain News Service: xi and 60; 163

National Gallery of Art
171

Courtesy of Pala Gems (palagems.com), Van Pelt Photographers
15

Project Gutenberg
Archaeological Essays by the Late Sir James Y. Simpson, 1872: 137

Shutterstock
© Jaroslav Moravcik: 2; © hobbit: 13t; © Byelikova Oksana: 89; © jsp: 100

Courtesy Sterling Publishing
13b

Courtesy Wellcome Library, London; www.wellcomeimages.org
3, 12, 80, 107, 164

Courtesy Wikimedia Foundation
The Yorck Project: *10.000 Meisterwerke der Malerei*: 27; W. and D. Downey: 28; British Library: 40; Cranach Digital Archive: 41; the David Collection: 65; Centro de Documentação D. João VI: 74

A Note on the Type

Gems & Crystals is composed largely in the "Scotch Roman" serif font
known as Miller Text, designed by award-winning, influential type
designer Matthew Carter, and released in 1997 by Font Bureau.
The sans serif font is Trade Gothic, originally designed
by Jackson Burke in 1948.

Interior production by 22MediaWorks, designed by Fabia Wargin.

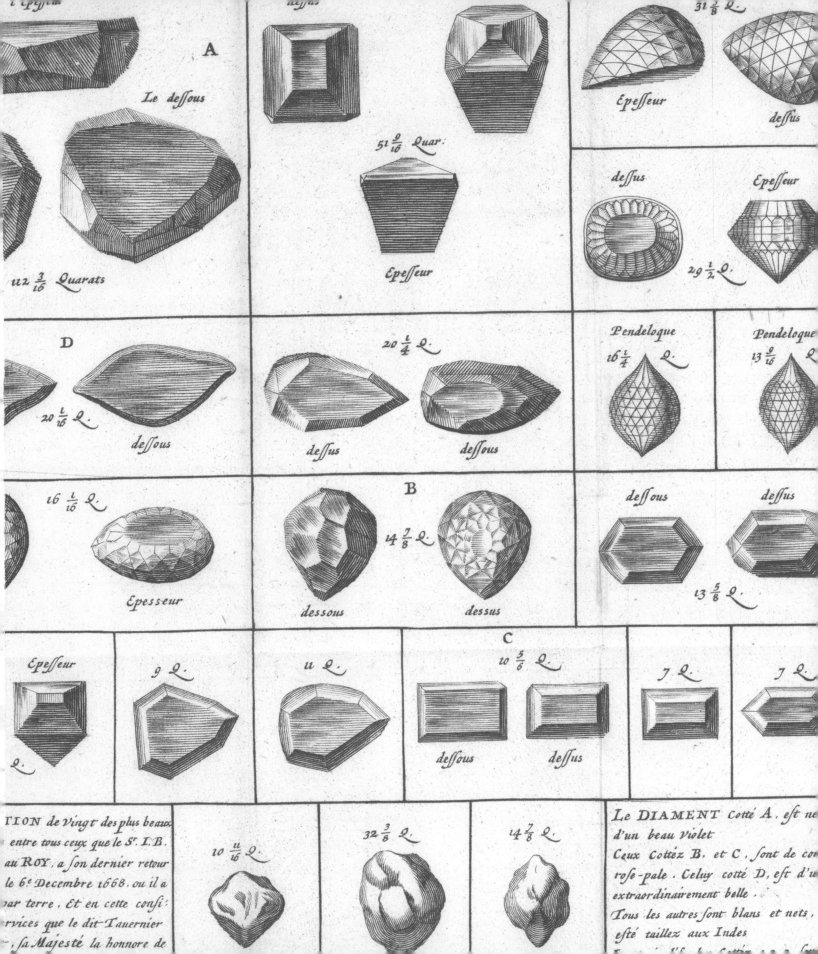

A

Le dessous

112 3/16 Quarats

51 9/16 Quar:

Epesseur

Epesseur

dessus

31 5/8 Q.

dessus

Epesseur

29 1/2 Q.

D

20 1/16 Q.

dessous

20 1/4 Q.

dessus   dessous

Pendeloque
16 1/4 Q.

Pendeloque
13 9/16 Q.

16 1/16 Q.

Epesseur

B

14 7/8 Q.

dessous   dessus

dessous   dessus

13 5/8 Q.

Epesseur

Q.

9 Q.

11 Q.

C

10 5/6 Q.

dessous   dessus

7 Q.   7 Q.

...TION de vingt des plus beaux
entre tous ceux que le Sr. I.B.
au ROY, a son dernier retour
le 6e Decembre 1668. ou il a
par terre, Et en cette consi-
vices que le dit Tauernier
, sa Majesté la honnore de

10 11/16 Q.

32 3/8 Q.

14 7/8 Q.

Le DIAMENT cotté A. est ne
d'un beau violet
Ceux Cottez B. et C, sont de cou
rose-pale. Celuy cotté D, est d'u
extraordinairement belle.
Tous les autres sont blans et nets,
esté taillez aux Indes